Norberto Cruz

Morphology and Histochemistry of the Midgut of Anticarsia gemmatalis

Norberto Cruz

Morphology and Histochemistry of the Midgut of Anticarsia gemmatalis

Analysis of the midgut of A. gemmatalis susceptible and resistant to AgMNPV fed with resistant genotypes

ScienciaScripts

Imprint
Any brand names and product names mentioned in this book are subject to trademark, brand or patent protection and are trademarks or registered trademarks of their respective holders. The use of brand names, product names, common names, trade names, product descriptions etc. even without a particular marking in this work is in no way to be construed to mean that such names may be regarded as unrestricted in respect of trademark and brand protection legislation and could thus be used by anyone.

Cover image: www.ingimage.com

This book is a translation from the original published under ISBN 978-613-9-68906-4.

Publisher:
Sciencia Scripts
is a trademark of
Dodo Books Indian Ocean Ltd. and OmniScriptum S.R.L publishing group

120 High Road, East Finchley, London, N2 9ED, United Kingdom
Str. Armeneasca 28/1, office 1, Chisinau MD-2012, Republic of Moldova, Europe
Printed at: see last page
ISBN: 978-620-8-24095-0

SUMMARY

DEDICATORY

To my parents Galdino Fernandes and Josefa Pereira for their teachings, affection and dedication.

To my siblings Màrcia, Màrcio and Vanilda for their great companionship.

To Ângela Falleiros, my advisor, friend and second mother.

ACKNOWLEDGMENTS

To God for his goodness and infinite greatness, for his protection, help and comfort in the most difficult moments of this journey.

To my advisor Ângela Maria Ferreira Falleiros for her guidance, trust, patience, friendship and teachings, both academic and in life.

To my friends from the master's program at the State University of Londrina, who have provided us with great moments of joy over the years.

To my great friends from undergrad André Scarpim, Evandro Prado, Èrica Rodrigues, Flàvia Mariano, Franciele Carvalho, Graziane Presotto, Luciano Chinen, Tiago Mendes, Rafael Spechotto, Willian Sakomura, Willian Takao, who, even from far away, always encouraged me to achieve my goals.

To my friends in Botucatu and Ilha Solteira who, directly or indirectly, helped me to carry out this work.

To the professors of the Agronomy Postgraduate Program at UEL for their teachings.

The technicians of the Histology Laboratory, Renata and Andréia, who contributed to this work.

Professor Sheila Michele Levy for her bibliographical contributions, help and advice, which have been invaluable over the years.

To my former classmates in Londrina (Màrio Hirano and Ricardo Okumura) and my current classmates (Tiago Baggio and Clàudio Sumida) who were and are my family here in Londrina.

To the new friends I made in Londrina - Everton Hiraoka, Edson Okumura, Ricardo Okumura, Màrio Hirano, Tiago Baggio, Jovenil, Gustavo, Claùdio Sumida, Sara Rovaris, Daniela Kelsener, Ernesto Lema, among others - who made life in Londrina a fun learning experience.

The Coordination for the Improvement of Higher Education Personnel (CAPES) for the scholarship granted during the course.

Embrapa Soja Londrina for providing the material to carry out the experiment.

To the other professors and staff of the Histology department.

To the secretary of the postgraduate section, Weda Westin, for her attention and friendship over the years.

SUMMARY

The soybean caterpillar, *Anticarsia gemmatalis* (Lepidoptera: Noctuidae) is the main pest of the soybean crop. To control this insect, techniques have been used such as biological control by the nucleopolyhedrovirus (*Ag* MNPV) and plant resistance with cultivars that have a complex of chemical substances (flavonoids) that can make them toxic to insects. The midgut is considered to be the place where the virus enters and where the defense substances present in resistant genotypes act. Flavonoids such as rutin and genistein can act to form fatty acids by hydrolyzing membrane phospholipids, altering the barriers and permeability of cell membranes. Apparently rutin and genistein inhibit digestion in insects and in *A. gemmatalis* larvae they have an antibiotic effect and reduce survival. In addition, studies have shown that rutin and genistein in *A. gemmatalis* larvae resistant to the *AgMNPV* virus have increased mortality compared to susceptible larvae. This study describes the morphology of the midgut of *AgMNPV-susceptible* and -resistant *A. gemmatalis* larvae using light microscopy and scanning electron microscopy, as well as the chemical composition of the cells after treatment with different soybean genotypes containing rutin and genistein at different concentrations.The midgut in both larval populations is composed of a peritrophic membrane, an epithelial layer with three cell types (columnar, goblet and regenerative) and a muscular layer. The epithelial layer shows morphological changes, especially in the columnar cells, with intense cytoplasmic vacuolization, the presence of cytoplasmic voids and an increase in protrusions released into the lumen. In susceptible larvae, these changes are evident in the PI 227687 genotype composed of R= 36.82 x 10^{-2} mg/g and G= 1.22 x 10^{-2} mg/g. In resistant larvae, the changes are evident in the genotypes IAC 100 composed of R= 9.72 x 10^{-2} mg/g and G= 1.42 x 10^{-2} mg/g and PI 227687 composed of R= 36.82 x 10^{-2} mg/g and G= 1.22 x 10^{-2} mg/g. After scanning microscopy analysis, the midgut is made up of three distinct regions: the anterior, middle and posterior regions, which are differentiated by the intensity of cytoplasmic projections. These projections increased in concentration in both larval populations after treatment with Dowling genotypes composed of R= 4.29 x 10^{-2} mg/g and PI 227687 composed of R= 36.82 x 10^{-2} mg/g and G= 1.22 x 10^{-2} mg/g. The histochemical techniques showed that the cells are composed of polysaccharides, glycogen, proteins and lipids. The highest concentrations of these compounds were found in the midgut of susceptible and resistant larvae after treatment with genotypes PI 229358, IAC 100 and PI 227687. Thus, the IAC 100 and PI 227687 genotypes are recommended for planting because they are resistant to the larvae and also because they promote the breakdown of resistance in resistant *A. gemmatalis* larvae.

Keywords: Soybean caterpillar. Flavonoids. Rutin. Genistein. Resistant plants

1 INTRODUCTION

Soya, *Glycine max* (L.) Merril, is a crop of great economic importance, generating foreign exchange in the agro-industrial sector, both in the marketing of grains and products and by-products derived from grains. Soya is also a great source of protein in human and animal feed (MISSĂO, 2006). Brazil is currently the second largest producer of soybeans on the world market, producing around 57.10 million tons of grain in the 2008/2009 harvest (CONAB, 2009).

During its cycle, soybeans can be attacked by agents that can affect its final production, mainly insect pests and diseases. Among the diseases, soybean rust or Asian rust (*Phakospsora pachyrhyzi*) is considered to be one of the most important and when left unchecked can cause major damage to the crop (DEGRANDE, 2006). With regard to insect pests that attack the leaves, the soybean caterpillar *Anticarsia gemmatalis* (Hübner 1818) (Lepidoptera: Noctuidae) is considered to be one of the main pests and when left unchecked it causes total destruction of the leaves.

In order to avoid reducing the damage caused by diseases and insects, the application of phytosanitary products is a very effective means of control. However, the indiscriminate use of these products has caused concern, especially with regard to ecological imbalance and the selection of resistant pest populations in the field, which is why more careful solutions have been sought, with Integrated Pest Management (IPM) guaranteeing the success of the crop.IPM consists of the harmonious combination of various control techniques applied in the field, the main objectives of which are to reduce the use of phytosanitary products (mainly chemical ones), seeking the least impact on the population of natural enemies, reducing environmental contamination and reducing production costs (GALLO et al., 2002; DEGRANDE, 2006).

Among the control techniques that are part of IPM, biological pest control through parasitoids, predators and entomopathogenic agents and the use of resistant plants and/or cultivars have been widely recommended for controlling the soybean caterpillar.

Generally, these resistant plants or cultivars have chemical substances in their leaf composition that make them toxic and repellent to insects (LARA, 1991). Among the flavonoids, rutin and genistein have been studied and studies indicate that when ingested by insects, they inhibit digestion (HOFFMANN-CAMPO, 1995).

In *A.gemmatalis* larvae susceptible and resistant to *AgMNPV*, these substances have shown adverse effects on their physiology and biology, and in some cases have increased the mortality of larvae resistant to *AgMNPV* (PIUBELLI, 2004).

As the digestive system is the main access route for food and plant protection products in general, it is expected that the ingestion of these substances will cause damage or alterations in the midgut of *A. gemmatalis* larvae susceptible and resistant to *AgMNPV*.

The objectives of this work were

- To verify the effects of the flavonoids rutin and genistein on the midgut *of A. gemmatalis* larvae susceptible and resistant to *AgMNPV*, using morphological techniques such as light and scanning electron microscopy.

- Identify the nature of the chemical substances of the epithelial cells present in the IM of these larvae, using histochemical methods.

- Compare the data obtained between LS and LR to those already described in the literature, and suggest through these results which genotype(s) cause(s) the greatest damage to the insect, i.e. the most resistant to attack by *A.gemmatalis*.

---

*This work was prepared in accordance with the "Norms of the Postgraduate Program in Agronomy at UEL for the presentation of Dissertations and Theses" - ABNT\NBR 6023: 2002 "with modifications".

2 LITERATURE REVIEW

2.1 SOY

Soybean, *Glycine max* (L.) Merril, is a legume belonging to the Fabaceae family, widely cultivated in several countries due to its great importance on the world market in recent decades. Its grains are one of the main sources of protein used in human and animal feed (EMBRAPA, 2005).

According to Missao (2006), a variety of important products and by-products are extracted from these beans, such as soybean meal, which is used as a food supplement in cattle, pig and poultry feed, and soybean oil, which is used in human nutrition, mainly in food preparation.

In addition to these products and by-products, soybeans and their derivatives are also used in industries, in the manufacture and production of biofuels, paints, cosmetics, pharmaceutical products, among others.

Today, cultivated soybeans are very different from their ancestors. Barbosa and Assumpçâo (2001) report that these plants were creeping, and their evolution began with the emergence of plants from natural crosses between two species of wild soybean, which were domesticated and improved by Chinese scientists.

This species of great economic importance was first mentioned between 2883 and 2838 BC, when it was considered a sacred grain, like rice, wheat, barley and millet. It was domesticated around the 11th century BC (EMBRAPA, 2006).

According to Costa and Manica (1996), the main center of origin of this culture is reported to be the Asian continent, in the region of ancient Manchuria in China, considered to be one of the oldest cultures. According to Missâo (2006), its expansion around the world began with English travelers and Chinese and Japanese immigrants.

In Brazil, the first historical record of soybeans was in 1882, in Bahia, where Gustavo Dutra cultivated the first germplasm of this crop brought from the United States, at the Bahia School of Agronomy. However, other historical records indicate that the "yellow" soybean was first cultivated at the Campinas Agronomic Station in 1891, to be tested as a fodder plant (COSTA; MANICA, 2006).

In 1900 and 1901, the São Paulo State Institute distributed the first seeds to São Paulo producers. During this same period, the first records of the crop appeared in Rio Grande do Sul and, in 1914, records were found of soya being planted for seed production and no longer as a fodder plant (COSTA; ALMEIDA, 2007).

In the state of Paranà, soybeans were first grown in the mid-1950s, when production was negligible until then. In 1953, after the great frost that destroyed the coffee plantations in the north, northwest, west and southwest regions, the crop was planted for commercial production, and since then the state has become the second largest producer in Brazil (EMBRAPA, 2006).

Currently, soybeans are of great economic importance on the world stage, with production reaching approximately 220 million tons and a planted area of 96.3 million hectares distributed all over the planet. The soybean agro-industrial complex is worth around US$ 215 billion a year, with the United States being the world's largest producer with an estimated production of 88 million tons, Brazil with 62 million tons, Argentina with 52 million tons and China with 16 million tons (CONAB, 2009).

South America has a cultivated area of approximately 41.5 million hectares and an estimated production of 95.3 million tons, in which Brazil participates with a cultivated area of 21.70 million hectares and an estimated production for the 2008/2009 harvest of 57.1 million tons. The largest national producers are the states of Mato Grosso, with a cultivated area of 5.8 million hectares and production of 17.963 million tons, and Paranà, with a cultivated area of 4.1 million hectares and production of 9.510 million tons (EMBRAPA, 2006; CONAB, 2009).

Soybeans develop between 100 and 160 days from emergence to maturity, divided into two phases: vegetative and reproductive. During these phases, until the crop is fully developed, it can suffer various injuries that can damage its production, especially problems with insect pests and diseases (VERNETTI, 1983; GALLO et al., 2002).

The diseases of economic interest that affect soybeans are those that attack the aerial parts, mainly oidium, mildew and Asian rust, where this disease can cause total crop loss if not properly controlled (DEGRANDE, 2006).

Among the complex of insect pests that attack soybeans, we can mention those that attack the leaves, such as the soybean caterpillar (*Anticarsia gemmatalis*), the false leafhopper (*Pseudoplusia includens*), the curling caterpillar (*Omiodes indicatus*), the cowfly (*Diabrotica speciosa*), and the whitefly (*Bemisia argentifoli*). In addition to these pests, we have those that attack the pods, such as the brown bug (*Euschistus heros*), green belly bug (*Dichelops melacantus*), small green bug (*Piezodorus guildinii*), green bug (*Nezara viridula*), edessa bug (*Edessa meditabunda*), acrostern bug (*Chinavia* spp.) and brown stink bug (*Thyanta perditor*), among others (GALLO et al., 2002; SOSA-GÓMEZ et al., 2006).

2.2 The Soybean Caterpillar

2.2.1 Description and Biology

Anticarsia gemmatalis (Hübner: 1818) is a pest belonging to the Order Lepidoptera and Family Noctuidae, with a complete cycle or holometabolus passing through different stages until its complete development, which usually lasts 47 days (egg: 3 days; larva: 15 days; pupa: 9 days and adult 20 days). It is a naturally occurring insect in tropical and subtropical areas, distributed in the Americas from Canada, the United States, Brazil and Argentina (FORD et al., 1975; GALLO et al., 2002). In Brazil, this pest occurs and causes severe damage from the south of Goiás to Rio Grande do Sul (PAN IZZI et al., 1977; GAZZONI et al., 1988; PANIZZI, 1990).

Its attacks can occur at different times according to each region of the country. In the northern region of Paranà, the attack begins in November, while in the other southern states the attack begins in December (HOFFMANN-CAMPO, 2000).

Although this insect is considered a key pest of the soybean crop, it also causes damage to other crops, such as cotton (DOUGLAS, 1930; HINDS; OSTERBERGER, 1931), alfalfa (ELLISOR; GRAHAN, 1937), pastures (WILLE, 1943), rice (TARRAGÓ et al., 1977) and wheat (MOSCARDI, 1979).

The eggs of *A. gemmatalis* are found deposited individually on the underside of leaves, and sometimes on stems and branches. They are oval or spherical in shape, with a flattened underside and vary in size from 0.5 to 2 mm in diameter. The coloration is initially light green, gradually darkening until close to hatching when they turn pink (WATSON, 1916) until they reach a reddish-brown color close to hatching (ELLISOR, 1942).

The newly hatched caterpillars have a chewing habit where they first feed on egg shells and then on young leaves. The small larvae (up to 10 mm long) are usually green in color and have four pairs of false legs, two vestigial and one terminal. As a result, they are palm-sized and are often confused with the *P. includens* caterpillar (GALLO et al., 2002; SOSA-GÓMEZ et al., 2006; DEGRANDE; VIVAN, 2009).

When they are more developed (greater than 15 mm in length), they lose the mede-palmo characteristic, become more active and their coloration can vary between green, reddish-brown or black, (the latter being evident only under conditions of high populations), as well as having five white longitudinal lines on the back (HOFFMAN-CAMPO et al, 2000; SOSA-GÓMEZ et al, 2006; DEGRANDE; VIVAN, 2009).

In general, *A. gemmatalis* goes through five larval stages, which last approximately 15 days and can reach up to 40 mm in length (SOSA-GÓMEZ, 2006). Reid (1975) reported that in populations of *A. gemmatalis* larvae subjected to temperatures above 18.3°C they had 5 to 6 instars, while at temperatures below 18.3°C these instars were 6 to 7. In similar studies, Silva (1981) observed that at temperatures varying between 20° and 35°C the larval period lasted between 34 and 13 days respectively.

When they approach the pre-pupa stage, the larvae are about 25 mm long and brown in color with a few longitudinal lines. This stage lasts approximately one day (WATSON, 1916; GUYTON, 1940). The newly formed pupa is light green in color, later turning dark brown and shiny. The larvae usually pupate in the soil, at a depth of up to 2 cm (HOFFMANN-CAMPO et al., 2000).

The pupal stage can last between 7 and 10 days depending on climatic conditions (HOFFMANN-CAMPO et al., 2006; DEGRANDE; VIVAN, 2009). Silva (1981) and Magrini (1996) studying temperature variation noted that the higher the temperature, the shorter the pupal stage. Silva and Parra (1986), studying the relationship between the photoperiod and the pupal stage, concluded that the longer the photoperiod, the longer the pupal stage.

The adult *A. gemmatalis* is a moth varying in color from grey, dark brown, beige, yellow or light blue with a wingspan of around 30 to 40 mm. At rest, the forewings cover the body, with an ever-present transverse line joining the tips of the first pair of wings (WATSON, 1915; HOFFMANN-CAMPO et al, 2000; GALLO et al., 2002; SOSA-GÓMEZ, 2006; DEGRANDE; VIVAN, 2009).

To differentiate between males and females in the adult stage, Greene (1974) described some characteristics to help with identification, in which males have elongated scales located on the femur of the prothoracic legs and on the tibia of the metathoracic legs, while females have these short, sparse scales.

During the day, moths are found in shady places at the base of plants, while the reproductive process takes place at night (GALLO et al., 2002). Greene et al. (1973) observed that the peak of mating between males and females occurs around 2 hours after sunset, gradually decreasing within the following 12 hours. Eggs are laid at night approximately 3 days after hatching (LEPPLA, 1976).

Each female can oviposit an average of around 1000 eggs (DEGRANDE; VIVIAN, 2008), although this average can vary according to ecological factors such as temperature, type of feeding (natural or artificial) and the phenological stage of the plant. SILVA (1981) observed that a temperature of 25°C is ideal for oviposition, where the average was 346.53 eggs/female, while at temperatures varying between 30° and 35°C, the average was between 26.6 and 35.1 eggs/female. The same author, studying the influence of diet or feeding, observed that insects fed an artificial diet had an average of 82.6 to 123.7 eggs/female, while those fed a natural diet of soybean leaves had an average of 345.56 eggs/female.

2.2.2 Damage

The larvae of *A. gemmatalis* go through five larval instars before reaching their adult form. In the first two instars, known as mede palmo, the larvae initially scrape the leaf parenchyma, causing small light spots on the leaves (HOFFMANN-CAMPO et al., 2000; GALLO et al., 2002).

From the third instar onwards, the larvae consume the leaf limb, leaving small holes in the leaves, thus causing a reduction in leaf area and photosynthetic rate (DEGRANDE; VIVAN; 2009).

This leaf consumption in the first three instars is considered small, but from the fourth to the fifth instar these larvae become more voracious and completely consume the leaves, damaging even the thinnest stems, where one larva can consume 100 to 150 cm^2 of leaf area, 95% of which occurs in the fourth to sixth larval instar. Thus, in high populations and if not controlled, these caterpillars can cause high levels of defoliation, causing yield losses, with damage that can reach 100% (HOFFMAN-CAMPO et al., 2000; GALLO et al., 2002; SOSA-GÓMEZ et al., 2006).

Depending on the time of the attack, the level of infestation and the environmental conditions, this leaf reduction can cause production losses (DEGRANDE; VIVAN, 2008).

The greatest damage can occur mainly during periods of shortage of rain (drought), where the caterpillars feed on both new and adult leaves, thus characterizing the so-called late attack which is considered the most damaging, as it occurs after flowering, where this period is essential and decisive for the filling of the pods (GAZZONI; YORINIORI, 1995).

Lourençâo et al. (2000) in trials with early and semi-early soybeans, under field conditions at stage R5 (grain filling), evaluated the percentage of leaf area cut (PAFC) by *A. gemmatalis*, and observed a reduction of 27.5% and 3% in the early genotypes *IAC* '5 and IAC 90-3395. *gemmatalis*, and observed that there was a reduction in leaf area in the early genotypes IAC '5 and IAC 90-3395, reducing it by 27.5 and 28.3%, and in the semi-early genotypes IAC 15, IAC 18 and IAC 93-2277 there was a reduction of 31.7 and 28.3%, where these values can lead to lower productivity.

2.2.3 Management and Control

The soybean crop is subject to attack by different pest species throughout its cycle. Even the presence of natural enemies such as predators, parasites and parasitoids, which partially reduce the population of these pests, is not enough to achieve total control and prevent damage to this crop. Thus, the use of phytosanitary products in soybean cultivation ends up being the most widely used way of controlling these pests (PRAÇA; SILVA NETO; MONNERAT, 2006).

In the case of pests, preventive application of phytosanitary products is not recommended because, in addition to interfering negatively with the population of natural enemies, causing an imbalance in the population of these insects, they can also increase production costs, which ends up being unfavorable for the agricultural producer.

Currently, the fight against pests in general is tending to gradually reduce the use of pesticides due to their side effects, being replaced by alternative control methods that take into account economic and ecological aspects for medium and long-term solutions. The strategy is a proposal for coexistence with pests, giving biological control a chance and recommending chemical control when the pest population reaches levels that cause greater damage than the costs of control. This proposal is based on ecological, economic and social knowledge, aiming to interfere as little as possible in the agro-ecosystem and is known as Integrated Pest Management (IPM), and can be characterized by the alignment of control methods with ecological, economic and social principles, aiming to interfere as little as possible in the agro-ecosystem.

Pest control should be based on the principles of Integrated Pest Management (IPM), which consists of the use of techniques that are employed harmoniously to solve a specific problem, i.e. its main objective is to reduce losses caused by pests in an effective, economically viable and ecologically compatible way (HOFFMANN CAMPO et al., 2006; DEGRANDE; VIVAN, 2009).

IPM in soybean cultivation aims to integrate traditional forms of control with alternative forms, rather than relying solely on control through the exclusive use of plant protection products (KOGAN ; HERZOG, 1980).

According to Costa and Almeida (2007) an average of 24% of production costs are spent on the purchase and application of insecticides and acaricides, so IPM aims to reduce the number of applications as well as the final costs, through the association of simple techniques such as: the use of resistant plants, soil preparation, crop rotation/suscession, sanitary measures (sanitary vacuum), biological control and the use of phytosanitary products that are selective to natural enemies (DEGRANDE; VIVAN, 2006).

Kogan and Herzog (1980) reported that one of the main tools for successful IPM is to monitor the different populations of insect pests at different stages of the plant's development, because it is through monitoring that we can determine when to start applying control methods.

This monitoring should be carried out through weekly sampling according to the species of insect pest. In the case of bedbugs and the soybean caterpillar, the white rag method is recommended, attached to two sticks, 1m long, which should be used on two rows of soybeans for caterpillars or one row of soybeans for bedbugs. This procedure should be repeated at various points in the field, taking the average of all the points sampled as the result (PRAÇA; SILVA NETO; MONNERAT, 2006, DEGRANDE; VIVAN, 2006).

In the case of attacks by defoliating caterpillars, such as *A.gemmatalis*, according to Praça; Silva Neto and Monnerat, (2006), the recommendation for starting control in the vegetative stage is 30% defoliation or 40 large caterpillars (larger than 1.5 cm) in 2 meters or 2 rows of soybeans beaten over the cloth, and in the reproductive stage this recommendation is 15% defoliation or 40 large caterpillars/ in 2 meters or 2 rows of soybeans beaten over the cloth.

In this way, the combination of all these mechanisms within IPM can lead to effective control of insect pests and thus provide the farmer with information on the best control alternative that can be taken for the problem in question (HOFFMANN-CAMPO et al., 2006).

2.3 Biological Control

Biological control in soybean cultivation consists of regulating the existing insect populations in this ecosystem, the main objective of which is to keep them within the maximum and minimum limits without causing alterations, i.e. within the economic damage threshold (STERN et al., 1965), which is defined as the smallest pest population that causes damage, while a relatively smaller insect population is defined as the economic damage threshold, which is regulated by its natural enemies such as predators, parasitic insects, parasitoids and entomopathogenic agents (GULLAN; CRANSTON, 2007).

In soybean cultivation, the population of natural enemies responsible for maintaining pest insect species are those belonging to the orders Hemiptera and Coleoptera (most predators), Diptera and Hymenoptera (parasitoids) and the entomopathogenic agents viruses, fungi and bacteria (HOFFMANN-CAMPO et al., 2000).

The group of natural enemies responsible for controlling *A..gemmatalis* include the parasites *Tricogramma* spp. which acts mainly on eggs, *Microcharops* a parasitoid of small larvae and *Patelloa similis* a parasitoid responsible for controlling large larvae (DEGRANDE; VIVAN, 2009).

The predators *Calosoma granulatun* (Coleoptera) and *Podisus* sp. (Hemiptera); the entomopathogenic agents *Baculovirus anticarsia* (virus), *Nomuraea rileyi* (fungus), and *Bacillus thuringiensis* (bacterium) are mainly responsible for controlling caterpillars (HOFFMANN-CAMPO et al., 2000; GALLO et al., 2002).

Among the entomopathogenic agents, *Bacillus thuringiensis* (*Bt)* is one of the main ones in terms of research and production, accounting for 90% of the biological products marketed worldwide, and has been used for more than three decades as a pest control agent (VALADARES-INGLIS et al., 1998).

Bt is a gram-positive, facultative aerobic bacterium that produces crystals that are toxic to the larvae of lepidopterans, dipterans and coleopterans. The effectiveness of this agent has led the biotechnology industry to introduce the producing genes into soybean plants, generating the so-called *Bt* plants, which are mainly aimed at controlling caterpillars (DEAN, 1984).

In general, the crystals produced when ingested by the larvae are affected by the intestinal pH and proteases, which solubilize the crystals and activate these toxins, which bind to receptors located in the epithelial tissue of the larva's midgut, causing the cell's osmotic balance to be broken (PRAÇA, 2004).

The epithelial cells swell and rupture, allowing the intestinal contents to leak into the insect's hemocele. As a result, the larva stops feeding, goes into general paralysis and dies from starvation or septicemia (MONERRAT; BRAVO, 2000).

In addition to the research and technology developed with *B. thuringiensis*, other natural enemies are reproduced and supplied to producers to help control the soybean caterpillar, such as *Trichogramma* spp. and *Baculovirus anticarsia* (EMBRAPA, 2005).

2.3.1 *Baculovirus Anticarsia* or *Anticarsa Gemmatalis* Multiple Nucleopolyhedrovirus (*AgMNNPV*)

Currently, a virus widely used as a bioinsecticide in Brazil is the Anticarsia gemmatalis nucleopolyhedrovirus (*AgMNPV*), belonging to the genus Nucleopolyhedrovirus (NPV), family Baculoviridae, specific to the soybean caterpillar. The pathology of this baculovirus mainly affects the larval stages of the insect and its main route of infection is through the ingestion of polyhedra, which cross the intestinal epithelium and multiply in the cells, progressing throughout the insect's body, resulting in a systemic infection. Also *in in vitro* systems, cells of A. gemmatalis and *Trichoplusia. ni*, this virus replicates generating a highly virulent infection (CASTRO et al., 2001).

AgMNPV is a naturally occurring virus that comprises the largest group of arthropod viruses and can infect around 700 species of arthropods (TANADA; KAYA, 1993). The majority have been isolated from the Lepidoptera order, however, it can be detected in other orders such as: Hymenoptera, Diptera, Coleoptera, Orthoptera, Neuroptera, Trichoptera (BILIMORIA, 1991; O'REILLY et al., 1992).

It has potential as a biological control agent for these insect pests in both agriculture and forestry (MARTIGNONI, 1984; MOSCARDI; SOSA-GOMEZ , 1992).

Baculovirus is composed of a circular double-stranded DNA surrounded by a protein capsid, which constitutes the infective unit of this virus (CASTRO et al., 1999). The Baculoviridae family comprises 2 genera: *Nucleopolyhedrovirus* (nuclear polyhedrosis virus - NPV), responsible for cell-to-cell transmission in the same individual or systemic infection and from insect to insect, and *Granulovirus* (granulosis virus - GV) responsible for transmission from insect to insect (GRANADOS; FEDERICI, 1986).

The main characteristic of these two genera is the occlusion protein bodies (IPB), which surround the viral particles, known as polyhedra for nuclear polyhedrosis viruses and granules for granulosis viruses. These protein crystals allow for the formulation of bioinsecticides that are easy to apply, representing savings and safety compared to traditional chemical products (MOSCARDI; SOUZA, 2002).

B. anticarsia was diagnosed for the first time in 1962, in Peru, in alfalfa crops, where it has been studied ever since. In Brazil, it was first reported in 1972 in soybean crops in Campinas, in the state of Sao Paulo, and later in other regions (SOSA-GÓMEZ, 1995).

From 1979 onwards, the Brazilian Agricultural Research Corporation (Embrapa Soja) developed a research program for the use of *AgMNPV*, isolating it from dead *A. gemmatalis* caterpillars in the northern region of Paranà, but it was during the 1980/81 and 1981/82 harvests that these products were tested on several properties in Paranà and Rio Grande do Sul, proving to be efficient in reducing pests (CASTRO et al., 1999).

Thus, in the following harvests, the use of the virus increased, reaching 1.4 million hectares of soybeans controlled with the virus, where 40% of this treated area was in the state of Paranà (MOSCARDI; SOUZA, 2002). However, some adverse factors limited the expansion of the use of the virus and contributed to a lack of interest on the part of private companies in producing and marketing this bioinsecticide. The high specificity of the *Baculovirus* (generally one species of pest and one particular crop), the high sensitivity of the viral polyhedra to solar radiation which deactivates the virus applied in 2 to 5 days, which would require the addition of sunscreens to the formulations, increasing the cost of manufacture, and the lack of acceptance by agricultural producers due to the relatively slow action of this agent on mortality and reducing damage to crops (MOSCARDI; SOUZA, 2002; DEGRANDE; VIVAN; 2009).

This relatively slow action can be directly attributed to the fact that the virus infection cycle in the insect begins when the larvae feed on host plants contaminated with (CPIs) and allow these bodies to penetrate the insect's digestive system. When the CPIs reach the lumen of the midgut, they are subjected to alkaline pH, which dissolves the polyhedrin, releasing the virions (nucleocapsid + envelope) into the intestinal lumen (GRANADOS; LAWLER, 1981; CASTRO et al., 1999). In this way, the epithelium of the midgut is identified as the first tissue to be infected by viruses during the infection process (KAWANISHI et al., 1972).

The virions cross the peritrophic membrane, considered the first protective barrier against pathogen action, fuse to the membranes of the epithelial cells of the midgut, replicate in the columnar cells and transmit the infection to other tissues (FLIPSEN et al., 1995; VOLKMAN, 1997; CASTRO et al., 1999; MOSCARDI; SOUZA, 2002).

Generally, within four days of ingestion and infection by the polyhedra, the larvae stop feeding and slow down their movement. In around seven or ten days, they die and have a whitish-yellow body color. As the days go by, they darken and rupture, releasing the virus into the environment which will serve as inoculum for new generations (GAZZONI et al., 1981; GAZZONI; YORINIORI, 1995; CASTRO et al., 1999).

Some studies with lepidopteran larvae infected with the nucleopolyhedrovirus have indicated that the infective process begins a few hours after the larvae ingest these polyhedra, and lead to a series of behavioral and morphological changes in these larvae that soon culminate in death. In the period between 48 and 78 hours, changes begin where there is a reduction in feeding and subsequent slowing of the insect's growth, which often does not change instar, in 80 to 100 hours there is discoloration of the insect's integument, and a few days after death there is rupture and release of the virus into the environment (GRANADOS; WILLIANS, 1986; VOLKMAN; KEDDIE, 1990).

2.4 Plant Resistance to Insects

Plant resistance to pests is yet another tool that helps in the effective control of insect pests and its main characteristic is its compatibility with other applicable control methods (chemical, biological and cultural), thus resulting in favorable synergistic effects (KOGAN, 1986).

In addition to the synergistic effect, plant resistance has other advantages in relation to its use in the field, such as: low cost, as this type of resistance does not add any financial value to the seed; low environmental impact, causing no adverse effects and being considered a soft technology in relation to the environment; ease of use, in which no additional knowledge about the pest or plant is required; persistence, where it acts permanently against low pest populations and reduces infestation in susceptible cultivars (GALLO et al., 2002).

Thus, the term plant resistance is directly attributed to inherited genetic qualities that result in one plant being less damaged than another (susceptible), where both are subject to the same environmental conditions, but do not possess the same resistance characteristics (PAINTER, 1951; GULLAN; CRANSTON, 2007).

These characteristics, in turn, can be influenced by the plant (age, physiological condition, part attacked), the insects (species, biotope, stage of development) and the environment (temperature, humidity, nutrients, etc.), determining the positive or negative manifestation of resistance (LARA, 1991).

According to Gullan and Cranston (2007), these resistance mechanisms are generally obtained through selective crosses for desirable traits. Painter (1951) proposed an empirical division in which the mechanisms are divided into three categories or levels: non-preference, antibiosis and tolerance.

Non-preference is when the cultivar is less used by the pest for feeding, oviposition or shelter compared to other cultivars under the same conditions. However, Kogan and Ortman (1978) coined the term antixenosis, which consists in part of the non-preference mechanism mentioned by Painter, but considers as additional or complementary factors the influence of secondary plant metabolites and morphological defenses, which together provoke a negative response in the insect during the host selection process.

Non-preference or antixenosis presents repellent and restricting factors such as plant chemicals, pubescence (covering of simple or glandular trichomes), surface waxes and leaf thickness or hardness, where the combination of all these factors prevents insect colonization (GULLAN; CRANSTON, 2007).

However, the combination of factors that confer plant resistance are not always favorable for different types of insect pests, where in bioassays carried out by Rossetto et al (1981), in which the authors studied the feeding preference of *Diabrotica speciosa* adults by offering different soybean cultivars that had characteristics of resistance and susceptibility to insect pests, they observed that the genotypes PI 227687, PI 229358 and PI 171451 were less damaged by *D. speciosa* adults. *speciosa*, characterizing non-preference.

In later studies carried out by Valle and Lourençâo (2002) analyzing the preference for oviposition by *Bemisia tabaci* Biotype B, with the same genotypes, they observed that the PI 227687 and PI 171451 genotypes were the most preferred for oviposition *by B. tabaci,* while the PI 229358 genotype had the lowest preference. This result can be attributed to the fact that the PI 227687 genotype has a favorable morphological characteristic for *B. tabaci:* the density of trichomes present on the leaves, which create a microclimate on the leaves favoring the nymphs until they fully develop.

The second level of resistance is linked to antibiosis, in which the insect normally feeds on the cultivar, but this exerts an adverse effect on its biology, causing mortality in the immature stage (often in the first instar), prolongation of the development period, reduction in size and weight, reduction in fecundity, fertility and oviposition period (GALLO et al., 2002).

This type of resistance is caused by factors such as: the presence of chemical substances, antimetabolites, enzymes, toxins, growth inhibitors, reduced nutrient levels, viscous exudates from glandular trichomes (hairs) and high concentrations of indigestible plant components such as silica (LARA, 1991; GALLO et al., 2002).

Fugi et al (2005) observed that the PI 229358 genotype caused high mortality of *A. gemmatalis* larvae, prolongation of the larval stage, lower pupal weights due to less feeding during the larval period, reduced oviposition and viability and a longer egg incubation period.

Tolerance is the third level of resistance, where only plant characteristics are involved and not the insect-plant interaction, in which the cultivar is little damaged compared to the others, without altering the behavior and biology of the insect, and these cultivars have the ability to tolerate the pest attack. In this way, the plant has the ability to tolerate the attack and outgrow or recover from defoliation or other damage caused by the insect feeding, leading to a reduction in the quantity/quality of its production (PAINTER, 1951; GULLAN; CRANSTON, 2007).

Generally, the resistance characteristics presented at the three levels are related to physical, chemical and morphological causes, in which the chemical causes are the most important, as they play a role in the choice of host, as well as in the amount of food that these insects will ingest (SRIVASTAVA et al., 1999).

These substances responsible for the chemical interactions between organisms are called allelochemicals, which are defined as secondary metabolites that affect survival, growth, behavior, fecundity or fertility, which are present in plants in the form of alkaloids, flavonoids, terpenoids and steroids (KUBO; HANKE, 1986). They are classified into two types: allomones that favor the sender (host plant acting as a defense substance) and cairomones that favor the receiver (in this case phytophagous insects) (GALLO et al., 2002).

According to Hoffmann-Campo (1995), the phenolic compounds present in soybean plants, in the form of flavonoids, are generally found in greater abundance in pest-resistant genotypes, where they can often be attractive, deterrent, repellent or toxic to pests, depending on their concentration (PIUBELII, 2004).

These flavonoids, in turn, make up a group of substances with a variety of biological activities. In the United States, three soybean introductions (PI 229358, PI 171451 and PI 227687) were discovered that offered resistance to multiple species of soybean pests, but without relating the cause of this resistance (CLARK et al., 1972; HATCHETT et al., 1977). In later studies, the phenolic compound or flavonoid rutin was identified in the leaves of wild soybean PI 227687, recognized for promoting plant defense against lepidopterans (HOFFMANN-CAMPO, 1995).

Thus, based on the results obtained by CLARK et al. (1972) with resistant plants, studies were carried out evaluating the resistance of F2 plants descended from the cross between commercial varieties and PI 229358, PI 171451 and PI 227687. The authors found that the progenies were less damaged by caterpillars and coleopterans and concluded that these plants transfer the resistance character to their progenies (REZENDE et al., 1980).

2.4.1 Flavonoids

The defense mechanisms of plants encompass a series of physical and morphological characteristics and a complex of chemical substances (phenolic compounds) that can make them unsuitable for consumption by insect pests (PIUBELLI, 2004).

In soybeans, these phenolic compounds or flavonoids are secondary metabolites found in various plant materials (HARBORNE, 1991).

Studies on the pharmacological effects of these substances in humans have been carried out, in which estrogenic action on the central nervous system, an increase in bone calcium, stimulation of mineralization and treatment of tumours, among others, have been demonstrated (SHILS et al., 2003); aesthetic effects have also been reported, such as in the treatment of sagging skin (BARNES et al., 2000). However, few studies have been carried out on their role in controlling insect pests. These compounds present in plants have different effects on herbivores, and their activation occurs through oxidation, leading to the precipitation of proteins. Thus, this activation can lead to an inhibition of digestion due to the formation of hydrogen bridges or the formation of covalent bonds with proteins and digestive enzymes (APPEL, 1993).

Appel (1993) reported that the hydroxyl radicals formed during the oxidation of phenolic compounds have a toxic action and are responsible for the disruption of membrane integrity and metabolic disorders in the intestinal epithelium. In Lepidoptera larvae, the covalent binding of chlorogenic acid to monomeric proteins inhibits the absorption of amino acids (FELTON et al., 1989; FELTON; DUFFEY, 1991).

The mode of action of flavonoids is related to the formation of fatty acids by the action of phospholipase A2, responsible for the hydrolysis of phospholipids present in cell membranes, with the release of arachidonic acid (KUBO; HANKE, 1986; LARA, 1991; SILVA et al., 2002; OMYLAGHA; GROTEWORLD, 2004; LOPES et al., 2005).

In this way, flavonoids can interact and penetrate the lipid bilayers of membranes, altering the barriers and permeability of cell membranes (LIMA et al., 1999; SILVA et al., 2001). Other functions include

related to acting on the expression of membrane receptors, as antibacterial and antiviral agents and inhibitors of proteolytic enzymes (TOLEDO et al., 2003).

To better understand the effects of flavonoids, Hoffmann-Field et al (1994) carried out food preference tests on *A. gemmatalis* larvae with various soybean genotypes BR 82-12547, IAC 74-2832, PI 227687, PI 229358 and PI 274454, and observed that these genotypes were the most rejected by the *A. gemmatalis* larvae. Similar studies and results were carried out and obtained by Rossetto et al (1981), on *D. speciosa*, in which the genotypes PI 227687, PI 171451 and PI 229358 were also rejected by the insect.

Subsequent studies using these genotypes identified the presence of seven flavonoids in PI 227687, one of the most widely used in the plant resistance to insects program, indicating greater toxicity of this cultivar to the insect. Among all the flavonoids identified, rutin (quercitin 3-O-rutinoside) and the isoflavonoid genistein (genistein 7-O-β-D-glucoside) are known to have adverse effects on lepidopteran larvae. These compounds have also been identified in extracts from the leaves of PI 274454 and other resistant genotypes (HOFFMANN-CAMPO, 1995, HOFFMANN-CAMPO et al.,

2001; PIUBELLI et al., 2005).

Rutin is a non-volatile compound that has a negative effect on the behavior and physiology of some species. Some results indicate that it breaks the resistance of the soybean caterpillar to baculovirus. According to Piubelli et al. (2005), this flavonoid inhibits digestion in herbivorous insects due to rupture of the cell membrane and metabolic disorders in the intestinal epithelium.

For Stamp and Skrobola (1993) this substance showed antibiotic and/or non-consumable effects on lepidopteran larvae such as *Manduca sexta*, *Heliothis virescens* (F.) (HOFFMANN-CAMPO, 1995) and *Trichoplusia ni* (Hübner) (HOFFMANN-CAMPO et al., 2001). In *A. gemmatalis*, rutin showed characteristic antibiosis effects, lengthening the larval cycle and reducing survival (GAZZONI et al., 1997).

Isoflavonoids, on the other hand, are phenolic chemical compounds widely distributed in plants, the highest concentrations of which can be found in legumes, particularly soybeans, both in leaves and pods (SETCVHELL, 1998).

Although genistein belongs to a different class of flavonoids, it has a wide range of biological properties, in which its combination with the flavonoid rutin causes changes in the biology of *A. gemmatalis*, prolonging the larval stage (GAZZONI; TUTIDA, 1996; FUGI et al., 2005).

The effects caused by these substances generally occur in the insects' alimentary canal. In *A. gemmatalis* larvae, most of the alterations occurred mainly in the midgut, where changes in the morphology of this alimentary canal were observed, as well as changes at the cellular level.

Salvador, (2008) studied the effects of phenolic compounds incorporated into artificial diets with the presence of casein, which can influence the activity of allelochemicals and additionally evaluated the effects of artificial diets containing different concentrations of rutin and genistein, as well as *fresh* soybean genotypes on biological aspects, as well as the mode of action of these foods in the midgut (MI) of *A. gemmatalis*.

The effect of rutin was only potentiated when the caterpillars were fed diets containing 7g of casein, i.e. half the amount needed for the insect to develop properly. Diets with the highest concentrations of rutin (R2) and the mixture R1+G1 and leaves of PI 227687 and 'IAC-100' were responsible for the greatest mortality, reduction in initial and pupal weights, lower weight gain and growth.

These treatments, especially R1+G1 and 'IAC-100', caused even greater morphological changes in the caterpillar's IM. Diets and genotypes containing only genistein did not show a marked anti-nutritional effect on the insect, nor any significant morphological changes. This author concluded that the interaction of substances R and G observed in soybean leaves, such as 'IAC-100' and PI 227687, acts to protect the plant from attack by *A. gemmatalis*, possibly by altering the columnar cells of the IM and suggested more complete studies.

2.5 *ANTICARSIA GEMMATALIS* DIGESTIVE TUBE

There is a great diversity of natural substrates in the environment that serve as food for insects, such as xylem sap (cercopid and cicada nymphs), vertebrate blood (bedbugs and female mosquitoes), dry wood (termites), bacteria and algae (leafhopper and trichopteran larvae), leaves (lepidopteran larvae), among others (GULLAN; CRANSTON, 2007).

For each different type of diet, insects have different mouthparts according to the type and texture of the food. In addition to this different mouth apparatus, they also have a digestive tract with different structure and functions (GALLO et al., 2002).

In insects where food is basically liquid, the digestive tract is generally long, narrow and coiled, as it must allow maximum contact with the liquid food. This alimentary canal has a special mechanism known as a filter chamber, which removes all excess water and has the function of concentrating nutrients and substances before digestion (GALLO et al., 2002).

On the other hand, insects that feed basically on solids have a digestive canal like a straight tube, long or short, with developed musculature and protection against abrasion or mechanical injury. This is the case of most lepidopteran larvae, grasshoppers and cockroaches (GULLAN; CRANSTON, 2007).

The alimentary canal of insects is divided into three regions, each of which performs a different function. These regions are duly separated by valves, which delimit the boundaries of each region: the cardiac valve (stomodeum and mesenterium) and pyloric valve (mesenterium and proctodeum) (CHAPMAN, 1998; GALLO et al., 2002).

The anterior region or stomodeum, is related to the ingestion, storage, grinding and transportation of food to the next region, the median or mesentery, where digestive enzymes are produced, where food absorption takes place and is considered the main access route for chemical and biological insecticides, and finally, the material remaining in the lumen of the digestive tract, together with the urine formed in the Malpighian tubules, reaches the posterior region or proctodeum, where water, mineral salts and other important molecules are absorbed before the feces are completely eliminated (CHAPMAN, 1998; LEVY et al., 2004).

Histological analysis of these three regions showed some differences between them: the anterior and posterior regions are practically identical, composed from the lumen to the periphery of an internal layer called the intima which has a cuticular structure, followed by a secretory epithelial layer which surrounds the intima and surrounding this structure is a muscular layer made up of longitudinal and circular muscles, which is surrounded by the peritoneal or basal layer. In the posterior region, in addition to these two layers of muscle (longitudinal and circular), it is formed and covered by a third layer of longitudinal muscle, these three layers of muscle have the function of favoring evacuation (GALLO et al., 2002; GULLAN; LEVY et al., 2004; CRANSTON, 2007).

In *A. gemmatalis* larvae, the midgut is composed from the lumen to the periphery of an acellular layer called the peritrophic membrane, followed by the epithelial layer and finally the surrounding muscular layer (LEVY et al., 2004; 2005; 2009).

The peritrophic membrane is an acellular layer, consisting of a network of chitin fibrils arranged in a protein-glycoprotein matrix (peritrophin) and a linear polymer N-acetyl-D-glucosamine, whose function is to protect the epithelium from abrasion by the intestinal contents, acting as a permeable barrier to digestive enzymes and digestion products (EISEMANN; BINNINGTON, 1994), as well as forming a barrier that prevents microorganisms from entering the body cavity and the hemocele (TERRA, 2001; WANG; GRANADOS, 2001).

This membrane in lepidopteran larvae is secreted by the columnar cells along the midgut and divides the intestinal lumen into two compartments, the ectoperitrophic space located between the membrane and the cellular epithelium in which digestive enzymes produced by the epithelium and substances that are being assimilated are found, and the endoperitrophic space, containing the food bolus in which food digestion takes place (HARPER; HOPKINS, 1997; HARPER; GRANADOS 1999; GALLO et al., 2002).

The peritrophic membrane is generally not found in insects that feed basically on liquids, such as Hemiptera and Thysanoptera, nor does it occur in adult insects of various orders, including lepidopterans (GULLAN; CRANSTON, 2007), where in some cases there is only the presence of a peritrophic gel that has the same functions as the peritrophic membrane, but does not have chitin in its composition (TERRA, 2001).

The epithelium of the midgut can be classified as simple or multiple. In lepidoptera larvae, this epithelium is pseudostratified columnar, formed by four distinct cell types: columnar, goblet, regenerative and endocrine cells (CHIANG et al., 1986, BINDER; BOWERS, 1994; BILLINGSLEY; LEHANE, 1996, LEVY et al., 2004). These cells are actively involved in the production and secretion of enzymes and the absorption of nutrients (CHAPMAN, 1998).

The different cell types that make up the epithelium have different morphological and functional characteristics. Columnar cells have an apical surface with microvilli, cytoplasmic projections and an elongated nucleus. They are abundant in the epithelium of the midgut and have various functions such as secreting digestive enzymes, the peritrophic membrane and absorbing assimilated digestive products (RICHARDS; DAVIES, 1994, LEHANE; BILLINGSLEY, 1996).

Goblet cells have invaginations of the apical plasma membrane, forming a goblet-like cavity, the inside of which is bordered by cytoplasmic projections apparently similar to microvilli containing mitochondria and a basal nucleus. These cells help the columnar cells with ion homeostasis and the absorption of metabolites (MOFFET et al., 1995; LEHANE; BILLINGSLEY, 1996; CHAPMAN, 1998).

The third cell type is the regenerative cells which are seen at the base of the epithelium, individually or in groups. These cells have scarce cytoplasm, a central nucleus and few organelles (DE PRIESTER, 1971; CAVALCANTE; CRUZ-LANDIM, 1999; LEHANE; BILLINGSLEY, 1996; LEVY et al., 2004).

According to Terra (1996), these cells can vary in both their morphology and quantity, depending on how close they are to moulting (ecdysis) and the life cycle.

These cells are mainly related to the process of epithelial renewal, where the eliminated cells are replaced by others developed from "nests" of regenerative cells (LEHANE; BILLINGSLEY, 1996; CHAPMAN, 1998; CAVALCANTE; CRUZ-LANDIM, 1999).

Endocrine cells are also located at the base of the epithelium, have a clear cytoplasm with a large number of cytoplasmic granules (EDWARDS; WEAVER, 2001) and are described in the intestinal epithelium of some insect orders such as some Lepidoptera (GÜL et al., 2001), Orthoptera (ENDO et al., 1982); Hymenoptera (CRUZ-LANDIM et al., 1996; NEVES et al., 2003), Odonata (ANDRIES, 1976) and Heteroptera (ANDRIES; BEAUVILLAIN, 1988). These cells are related to hormonal control, peristaltic movements, digestion, diuresis, development and reproduction (BROWN et al., 1985; PRATT et al., 1989; WOODHEAD et al., 1989; SCOOFS et al., 1993, COAST; KAY, 1994; PATEL et al., 1995).

According to Iwanaga et al. (1981), these cells have functional similarities to the endocrine cells found in vertebrates, i.e. they are neurosecretory cells that synthesize and release large quantities of polypeptide hormones (MARTOJA; BALLAN-DUFRANÇAIS, 1984).

These secretion products are related to the differentiation of regenerative cells and control the secretion of digestive enzymes produced by these midgut epithelial cells (BILLINGSLEY; DOWNE, 1986; CAVALCANTE; CRUZ-LANDIM, 1999).

The epithelium is supported by a continuous basement membrane of connective origin and by the muscular layer, which is made up of longitudinal and circular muscles. The overlapping of the muscles differs in relation to the anterior intestine, i.e. the circular muscles appear internally positioned, while the longitudinal muscles are externally allocated (CONTE, 1985; LANE et al., 1996; GALLO et al., 2002).

According to Gallo et al. (2002) these muscles have the function of keeping food moving inside the ventricle, favoring digestion and promoting peristaltic movements, taking the food bolus towards the proctodeum or posterior intestine.

Considering the architecture of the intestinal epithelium of lepidopteran larvae, Levy et al. (2009), in comparative morphometric studies between populations of *A. gemmatalis* larvae resistant and susceptible to AgMNPV, observed that the epithelium of the midgut showed differences between the two larval populations. In resistant larvae, these epithelial cells had higher morphometric values than susceptible larvae and they suggested that this could be related to the resistance of these larvae to *AgMNPV*.

2.6 Soybean Caterpillar Resistance to AgMNPV

Insects generally have a great ability to adapt to different environments, even if these have characteristics that are totally adverse to their survival. Generally, insects create morphological mechanisms (such as a more resistant exoskeleton) and genetic mechanisms (resistant populations) against these adversities.

As insect pests of various crops are basically controlled through the use of phytosanitary products, it is often the case that applying a product just once is not enough to reduce the pest population in the crop. Applying the same product more than once, and at a higher dose to achieve the ideal level of pest control that does not cause economic damage to the crop, can lead to the appearance of pest insect populations that are resistant to a given plant protection product (GALLO et al., 2002; GULLAN; CRANSTON, 2007). The main resistance mechanisms of insects in this case are biochemical, physiological and behavioral (DEGRANDE; VIVAN; 2008).

The term resistance is applied by Lara (1991) to characterize previously susceptible insect species whose population can no longer be controlled by a given product at the normally recommended dose, or at no dose at all, in which case this characteristic is transferred to the next generations.

The factors affecting insect resistance to insecticides are related to: (1) genetic factors - number of resistant alleles, frequency and intensity of resistance, dominance of resistant alleles, adaptive value of resistant individuals; (2) bio-ecological factors - such as the number of generations per year, reproduction rate, mode of reproduction, mobility of the species, feeding habits, the presence of refuges for the susceptible population and the presence of effective natural enemies of the pest, (3) operational factors - characteristics of the chemical compound (group, persistence, selectivity and formulation), and application characteristics (level of control, method of application, strategy for the use of chemical products (GALLO et al., 2002; GULLAN; CRANSTON, 2007).

Most cases of resistance in the field are linked to chemical products, with no evidence to date of populations resistant to biological products.

Examples of pest-resistant insect populations include the tobacco whitefly *Bemisia tabaci* biotype B, the crucifer moth *Plutella xylostela* and the potato beetle *Leptinotarsa decemlineata*, which are resistant to the vast majority of chemical insecticides on the market (GULLAN; CRANSTON, 2007).

In soybeans, Sosa-Gómez (1995), studying the active ingredient endosulfam in the control of the brown stink bug *Euchistus heros*, found populations resistant to this active ingredient.

This fact can be associated with two mechanisms related to resistance: a physiological mechanism, where the action of enzymes, mainly mixed-function oxidases, reduces the penetration and transport of insecticide molecules to the central nervous system; and a second mechanism related to the behavioral factor, in which the irritating and repellent effect of the product causes individuals to change their behavior and avoid treated areas (DEGRANDE; VIVAN, 2009).

This development of insecticide resistance is linked to a number of factors such as frequent application of the insecticide, an increase in the dosage of the product, the use of improper mixtures of products and substitution with another generally more toxic product (GALLO et al., 2002).

Abot et al (1995), carrying out laboratory experiments involving selection pressure on *A. gemmatalis* larvae with the *AgMNPV* virus, observed rapid selection for resistance to this entomopathogen. The emergence of resistant larvae led these researchers to question the mechanisms of resistance developed by the insect. Some studies indicate that this resistance may be attributed to differential defense mechanisms in the midgut (MOSCARDI , 1986; ABOT et al., 1996). This fact was corroborated by Levy et al. (2005), in comparative studies between *A. gemmatalis* larvae susceptible and resistant to *AgMNPV*. They observed that the population of resistant larvae showed a higher concentration of regenerative cells and columnar cells with a large number of cytoplasmic protrusions.

On the other hand, in pest control programs to reduce the resistance of *A. gemmatalis* larvae, chemical substances found in extracts of soybean leaves have been used, which promote the mortality of larvae resistant to the virus compared to susceptible larvae.

As the midgut of insects is the main route of access and action of these substances, there is a need to study the alterations caused by them. This work aims to correlate the post-ingestive effects of the flavonoid rutin and the isoflavonoid genistein, both alone and in combination, in the midgut of *A. gemmatalis* larvae susceptible and resistant to *AgMNPV* and to relate the effects of these substances present in the genotypes in breaking the resistance of these larvae, as well as indicating which soybean genotype promotes the greatest changes in the midgut of both larval populations.

REFERENCES[1]

ABOT, A.R.; et al. Susceptibility of populations of *Anticarsia gemmatalis* (Lepidoptera: Noctuidae) from Brazil and the United States to a nuclear polyhedrovis virus. **Journal of Entomological Science,** Georgia, v.30, p.62-69, 1995.

ABOT, A.R.; et al. Development of resistance by *Anticarsia gemmatalis* from Brazil and the United States to a nuclear polyhedrosis virus under laboratory selection pressure. **Biological Control**, Montreal, v.7, n. 1, p.126-130, aug. 1996.

ANDRIES, J.C. Presence of two endocrine cell types and one exocrine type in the mesenteron of the larva of *Aeshna cyanea* Muller (Odonata: Aeshnidae). **International Journal of Insect Morphology and Embryology**, Ouxford, v.5, n.6, p. 393-407, Oct. 1976.

ANDRES, J.C.; BEAUVILLAIN, J.C. Ultrastructural study of cholecystokinin-like immunoreactivity in endocrine cells of the insect midgut of *Nepa cinerea* (Insecta, Heteroptera): ultrastructure and genesis. **Biology of the Cell**, Paris, v.46, p.195-202, 1988.

APPEL, H.M. Phenolics in ecological interactions: the importance of oxidation. **Journal of Chemical**

1 Bibliographical references according to the rules of the Master's program in Agronomy at the State University of Londrina - UEL, 2008.
BRAZILIAN ASSOCIATION OF TECHNICAL STANDARDS. NBR **6023:** references, information, documentation and preparation. Rio de Janeiro, 2002

Ecology, California, v.19, n.7, p.1521-1551, jul.1993.

BARBOSA, M.Z.; ASSUMPÇÂO, R. Territorial occupation of soybean production and agroindustry in Brazil in the 80s and 90s. **Informaçôes Econômicas**, Sao Paulo, v.31, n.11, p. 7-17, nov. 2001.

BARNES, S. et al. Beyond ERa and ERβ: Estrogen receptor binding is only part of the isoflavone story. **The Journal of Nutrition**, Birmingham, v.1, n.30, p.656S-657S, 2000.

BILIMORIA, S.L. The biology of nuclear polyhedrosis viruses. In: KURSTAK, E. (Ed.). **Viruses of invertebrates**. New York: Marcel Dekker, 1991. p.1-72.

BILLINGSLEY, P.F.; DOWNE, A.E.R. Non digestive cell types in the midgut epithelium of *Rhodnius prolixus* (Hempitera: Reduviidae). **Journal of Medical Entomology**. Florida, v.23, p.212-216, 1986.

BILLINGSLEY, P.F.; LEHANE, M.J. Structure and ultrastructure of the insect midgut.

In: LEHANE, M.J., BILLINGSLEY, P.F. (Eds). **Biology of the Insect Midgut**.

London: Chapman; Hall, 1996, p.3-30.

BINDER, B.F.; BOWERS, W.S. Precocene II-induced changes in the anatomy of midgut goblet cells of last instar larvae *of Helicoverpa zea* (Boddie) (Lepidoptera: Noctuidae). **International Journal of Insect Morphology and Embryology**. Oxford v. 23, n.2, p.127-134, Apr. 1994.

BROWN, M.R., RAIKEL, A.S., LEA, A.D. Ultrastructure of midgut endocrine cells in the adult mosquito, *Aedes aegypti.* **Tissue and Cell**, Ouxford, v.17, n.5, p. 709-722, aug. 1985.

CASTRO, M.E.B.,et al. Molecular biology of baculoviruses and their use in pest control in Brazil. **Pesquisa Agropecuaria Brasileira**, Brasilia, v.34, n. 10, p. 17331761, Oct. 1999.

CAVALCANTE, V.M.; CRUZ-LANDIM, C. Types of cells present in the midgut of the insects: a review. **Naturalia**, Rio Claro, v.24, p.19-40, 1999.

CHAPMAN, R.F. **The Insects: structure and function**. 4.ed. Cambridge: Havard University Press, 1998.

CHIANG, A.S.; YEN, D.F.; PENG, W.K. Defense reaction of midgut epithelial cells in the rice moth larva (*Corcyra cephalonica*) infected with *Bacillus thuringiensis*.

Journal Invertebrate Pathology,Netherlands, v.47,n.3, p.333-339, May 1986.

CLARK, W.J.; et al. Resistance of certain soybean cultivars to bean leaf beetle blister beetle, and bollworm. **Journal Economic Entomology**, Baltimore, v.65, p. 16691672,1972.

COAST, G.M., KAY, I. The effects of *Acheta* diuretic peptide on isolated Malpighian tubules from the house cricket, *Acheta domesticus*. **The Journal of Experimental Bio logy**, London, v.187, n.1, p.225-243, feb. 1994.

COMPANHIA NACIONAL DE ABASTECIMENTO. **Agricultural Indicators.** Available at: http://www.conab.gov.br/conabweb/IA-agos09.pdf. Accessed on August 6, 2009.

CONTE, H. **Morphology of the digestive tract *of Diatrea saccharalis* (Fabricius, 1794) (Lepidoptera, Pyralidae) during larval development, anatomical and histological study.** 1985. Dissertation (Master's Degree in Biology) Institute of Biosciences, Paulista State University, Rio Claro.

COSTA, A.J.D.; ALMEIDA, M.J.M. The agro-industrial complex in Paraná: transgenic soy versus conventional soy. In: ECOPAR, n.5, 2007, Curitiba. **Proceedings...** Curitiba: UFPR, 2007 p.1-20.

COSTA, J. A.; MANICA, I. **Soybean Culture.** Porto Alegre: Ceres, 1996.

CRUZ-LANDIM, C. et al. Ultrastructural aspects of epithelial renewal in the midgut of adult worker bees (Hymenoptera, Apidae). **Journal of Computational Biology**, New Rochelle, v.1, p.29-40, 1996.

DEAN, D.H. Biochemical genetics of the bacterial insect-control agent *Bacillus thuringiensis*: basic principles and prospect for genetic engineering. **Biotechnology and Genetic Engineering Reviews**, Andover, v.2, p.341-363, 1984.

DEGRANDE, P.E.; VIVAN, L.M. Pragas da Soja. In: **Soybean Research Bulletin**. Rondonópolis: Fundaçâo MT, 2006. p.153-169.

DEGRANDE, P.E.; VIVAN. L.M. Soybean pests. In: ________. **Technology and Production**: Soybeans and Corn 2008/2009. Rondonópolis: Fundaçâo MT, 2009. p.73-108.

DE PRIESTER, W. Ultrastructure of the midgut epithelial cells in the fly *Callifora erythrocephala*. **Journal of Ultrastructure Research.** New York v.36, n.5, p. 783805, set. 1971.

DOUGLAS, N. A. The velvetbean caterpillar as a pest of soybeans in southern Louisiana and Texas. **Journal of Economic Entomology**, Baltimore, v. 23, n.4, p.683-690, aug. 1930.

EDWARDS, J.P.; WEAVER, R.J. **Endocrine interactions of insect parasites and pathogens.** Trowbridge: The Crowell Press, 2001.

EISEMANN, C.H.; BINNINGTON, K.C. The peritrophic membrane : its formation, structure, chemical composition and permeability in relation to vaccination against ectoparasitic arthropods. **International Journal for Parasitology**, England, v.24, n.1, p.15-26, feb. 1994.

ELLISOR, L. O.; GRAHAM, L. T. A recent pest of alfalfa. **Journal of Economic Entomology**, Baltimore, v. 30, n. 2, p. 278-280, 1937.

ELLISOR, L. O. **Notes on the biology and control of the velvetbean caterpillar, *Anticarsia gemmatalis* Hbn.**Louisiana: Louisiana Agricultural Experimental Station Bulletin, 1942. v.350

EMBRAPA. **Production Systems: Technology and soybean production in central Brazil 2006.** 1.ed. Londrina: Embrapa-CNPSo, 2006.

EMBRAPA SOJA. **Soybean Production Technology in Paranà 2004**. Available at: <http://www.cnpso.embrapa.br/producaosojaPR/SojanoBrasil.htm> Accessed on: April 23, 2009.

ENDO, Y. et al. Fine structure of developmental endocrine cells and columnar cells in the cockroach midgut. **Biomedical Research**, Japan, v. 3, n.6. p. 637-644, 1982.

FELTON, G.W. et al. Activation of plant foliar oxidases by insect feeding reduces nutritive quality of foliage for noctuid herbivores. **Journal of Chemical Ecology**, California, v.15, n.12, p.2667-2694, 1989.

FELTON, G.W.; DUFFEY, S.S. Enzimatic antinutritive defenses of the tomato plant against insects.In: HEDIN, P. (Ed.) **Naturally occurring pest bioregulators.**

Washington: American Chemical Society, 1991. p.166-197.

FLIPSEN, J.T.M.; et al. Passage of *Autographa californica* nuclear polyhedrosis virus through the midgut epithelium of *Spodoptera exigua* larvae. **Virology**, New York v.208, p.328-335, 1995.

FORD, B. J. et al. **The literature of arthropods associated with soybean, a bibliography of the velvetbean caterpillar *Anticarsia gemmatalis* Hübner (Lepidoptera: Noctuidae)**. Urbana: Illinois National History Survey, 1975.

FUGI, C.G.Q.; et al. Biology of *Anticarsia gemmatalis* on Soybean Genotypes With

Different Degress of Resistance to Insects. **Scientia Agricola**, Piracicaba, v.62, n.1, p.31-35, jan./feb. 2005.

GALLO, D.; et al. **Entomologia agricola**. Piracicaba: FEALQ, 2002. v.10.

GAZZONI, D.L. et al., **Soybean pest management.** Londrina: Embrapa-CNPSo, 1981.

GAZZONI, D. L.; et al. **Soybean pest management**. Londrina: Embrapa-CNPSo, 1988.

GAZZONI, D.L.; YORINIORI, J.T. **Manual de identificaçâo de pragas e doenças da soja.** Brasilia: EMBRAPA, 1995.

GAZZONI, D.L.; TUTIDA, F. Effect of resistant and susceptible genotypes on the biology of the soybean caterpillar (*Anticarsia gemmatalis* Hübner). **Pesquisa Agropecuâria Brasileira**, Brasilia , v.31, n.10, p. 709-714, Oct. 1996.

GAZZONI, D.L. HÜLSMEYER,A.; HOFFMANN-CAMPO, C.B. Effect of different doses of rutin and quercetin on the biology of *Anticarsia gemmatalis* Hübner, 1818 (Lep., Noctuidae). **Revista Agropecuâria Brasileira**, Brasilia, v.32, n.7, p. 673-681, jun./jul. 1997.

GRANADOS, R.R.; LAWLER, K.A. In vivo pathway of *Autographa californica* baculovirus invasion and infection. **Virology**, New York, v.108, n.30, p.297-308, jan. 1981.

GRANADOS, R.R.; FEDERICI, B.A. (Eds.). **The biology of baculoviruses**. Boca Raton: CRC, 1986. v.1.

GRANADOS, R.R.; WILLIAMS, K.A. In vivo infection and replication of baculoviruses. In: GRANADOS, R.R.; FEDERICI, B.A. (Eds.). **The biology of baculoviruses**. Boca Raton: CRC,

1986, p.89-108 v.1.

GREENE, G. L.;et al. Mating and oviposition of the velvetbean caterpillar in soybeans. **Environmental Entomology**, College Park, v. 2, n. 6, p. 1113-1115, 1973.

GREENE, G. L. Sexual dimorphism of *Anticarsia gemmatalis* leg scales. **Florida Entomologist**, Gainsville, v. 57, n. 3, p. 280, 1974.

GULLAN, P.J.; CRANSTON, P.S. **Os insetos um resumo de entomologia**. 3 ed. Sao Paulo: Roca, 2007.

GÜL, N. et al. A study on endocrine cells in the midgut of *Agrotis segetum* (Denn and Schiff.) (Lepidoptera: Noctuidae). **Turkish of Journal Zoology**, Ankara, v. 25, p. 193-197, Oct. 2001.

GUYTON, F. E. An outbreak of the velvetbean caterpillar in Alabama with data on control. **Journal of Economic Entomology**, Baltimore, v. 33, n. 4, p. 635-639, 1940.

HARBONE, J.B. Flavonoid pigments. In: ROSENTHAL, G.A.; BEREMBAUM, M.R. (Eds.). **Herbivores and their interaction with secondary metabolites**. London: Academic Press, 1991, p.389-429.

HARPER, M.S.; HOPKINS, T.L.. Peritrophic membrane structure and secretion in European corn borer larvae (*Ostrinia nubilalis*). **Tissue and Cell,** Ouxford, v.29, n.2, p.463-475, aug. 1997.

HARPER, M.S.; GRANADOS,R.R. Peritrofic membrane structure and formation of larval *Trichoplusia ni* with na investigation on the secretion patterns of a PM mucin. **Tissue and Cell**,Ouxford, v.31, n.2, p.202-211, May, 1999.

HATCHETT, J.H.; BELAND, G.L.; HARTWING, E.E. Leaf feeding resistance to bollworm and tobacco budworm in three soybean plant introductions. **Crop Science**, Stanford, v.16, p. 277-280, 1977.

HINDS, W. E.; OSTERBERGER, B. A. The soybean caterpillar in Louisiana. **Journal of Economic Entomology**, Baltimore, v. 24, n.6, p.1168-1173, 1931.

HOFFMANN-CAMPO,C.B.; et al. Mechanisms of resistance of soybean genotypes: non-preference test for *Anticarsia gemmatalis* Hübner, 1818 (LepzNoctuidae). **Pesquisa Agropecuaria Brasileira**, Brasilia, v.29, n.4, p. 513-519, abr. 1994.

HOFFMANN-CAMPO, C.B.; **Role of the flavonoids in the natural resistance of soyabean to *Heliothis virescens* (F.) and *Trichoplusia ni* (Hübner)**. 1995.

Dissertation (PhD in Botany) - 165 p. The University of Reading, Reading.

HOFFMANN-CAMPO, C.B.; et al. **Soybean pests in Brazil and their integrated management.** Londrina: EMBRAPA-CNPSo, 2000.

HOFFMANN-CAMPO, C.B.; et al. Pre-ingestive and and post-ingestive effects of soybean extracts and rutin on *Trichoplusia ni* growth. **Entomologia Experimentalis et Applicata**, Netherlands, v.98, n.2, p.181-194, feb. 2001.

HOFFMANN-CAMPO, C.B.; et al. Detrimental effect of rutin on *Anticarsia gemmatalis*. **Pesquisa Agropecuaria Brasileira**, Brasilia, v.41, n.10, p. 1453-1459, Oct. 2006.

IWANAGA, T.; et al. Immunohistochemical demonstration of PP⁻, somatostatin- enteroglucagon and VIP⁻ like immunoreactivites in the cock-roach midgut.

Biomedical Research, Japan, v.2, p.202-207, 1981.

KAWANISHI, C.Y.; et al. Entry of insect virus *in vivo* by fusion of viral envelope and microvilli membrane. **Journal of Invertebrate Pathology**, Netherlands, v.20, p.104108, jul. 1972.

KOGAN, M.; ORTMAN, E.E. Antixenosis - a new term proposed to replace Painter's "Non-preference" modality of resistance. **Bulletin of the Entomological Society America**, Lanham, v.24, p-175-176, 1978.

KOGAN, M.; HERGZOG, D.C. **Sampling methods in soybean entomology.** New York: Springer, 1980.

KOGAN, M. Natural chemicals in plant resistance to insects. **Journal Research**, lowa State v.60, p.501-527, 1986.

KUBO, I.; HANKE, F.G. Chemical methods for isolating and identifying phytochemicals biologically active in insects. In: MILLER, J.R.; MILLER, T.A. (Eds) **Insects plants interactions**. New York: Spring Verlog. 1986 p.225-249.

LANE, N.J.; DALLAI, R.; ASHHURST, D.E. Structural macromolecules of the cell membranes and the extracellular matrices of the insect midgut, In: LEHANE, M.J.; BILLINGSLEY, P.F. (Eds.) **Biology of the Insect Midgut**. London: Chapman; Hall, 1996. p.115-150.

LARA, F.M. **Principios de resistência de plantas a insetos**. Sao Paulo: icone, 1991,

LEPPLA, N. C. Circadian rhytms of locomotion and reproductive behavior in adult velvetbean caterpillar. **Annals of the Entomological Society of America**, Columbus, v. 69, n. 1, p. 45-48, 1976.

LEVY, S.M. et al. The larval midgut of *Anticarsia gemmatalis* (Hübner) (Lepidoptera: Noctuidae): light and electron microscopy studies of the epithelial cells. **Brazilian Journal of Biology**, Sao Paulo, v.64, n. 3b, p.1-6, aug. 2004.

LEVY,S.M. et al. Morphology of the midgut *of Anticarsia gemmatalis,* susceptible and resistant to nuclear polyhedrosis virus (*AgMNPV*), in control and virus-infected larvae. In: SIMPÒSIO DE CONTROLE BIOLÒGICO, 4., 2005, Recife. **Abstracts...** Recife: FIOCRUZ, 2005. p.188.

LEVY, S.M.;et al. A morphometric study of the midgut in resistant and non-resistant *Anticarsia gemmatalis* (Hübner) (Lepidoptera: Noctuidae) larvae to its nucleopolyhedrovirus (AgMNPV).

Journal of Invertebrate Pathology, Netherlands, v.101, n.1, p.17-22, Apr. 2009.

LIMA, L.R.P. et al. Determination of lipase activity in the presence of morin, naringenin, naringin and rutin. **Revista de Ciência e Agrotecnologia**, Lavras, v. 23, n. 3, p. 626-631, jul./set. 1999.

LOPES, R.M. et al. Effect of different doses of rutin on the concentrations of albumin, creatine, urea and uric acid in the serum of rabbits. **Revista Brasileira de Farmacia**, Rio de Janeiro, v. 86, n.1, p. 35-38, jan. 2005.

LOURENÇÂO, A.L.; et al. Evaluation of damage caused by bedbugs and caterpillars in early and semi-early cycle soybean genotypes. **Pesquisa Agropecuària Brasileira**, Brasilia, v.35, n.5, p. 879-886, May 2000.

MAGRINI, E. A.; SILVEIRA NETO, S.; PARRA, J. R. P.; BOTELHO, P.S.M. Biologia e exigências *térmicas de Anticarsia gemmatalis* Hübner (Lepidóptera: Noctuidae) em laboratório. **Anais da Sociedade Entomológica do Brasil**, Londrina, v. 25, n. 3, p. 513-519, dec. 1996.

MARTIGNONI, M.E. Baculovirus: an attractive biological alternative. In: GARNER,W.Y.; HARVEY JUNIOR, J. (Eds.). **Chemical and biological controls in forestry.** Washington, D.C.: American Chemical Society, 1984. p.5567.

MARTOJA, R., BALLAN-DUFRANÇAIS, C. The ultrastructure of the digestive and excretory organs. IN: KING,R.C.; AKAI, H. (Eds.). **Insect ultrastructure**. New York: Plenum Press, 1984. v.2, p.119-261.

MISSÃO, M.R. Soy: origin, classification, use and a comprehensive view of the market. **Maringa Management: Ciências Empresariais**, Maringà, v.3, n.1, p 7-15, jan./jun. 2006.

MOFETT, F.; KOCH, A. Electrophysiology of K^+ transport by midgut epithelium of lepidopteran insect larvae. **The Journal of Experimental Biology**, London, v.198, n.10, p.2115-2125,Oct. 1995.

MONNERAT, R.G.; BRAVO, A. Bioinsecticidal proteins produced by *Bacillus thuringiensis* bacteria mode of action and resistance. In: MELO, L.S.; AZEVEDO, J.L. (Ed). **Controle biològico**, Jaguariùna Embrapa-CNPMA, 2000, v.3, p. 163-200.

MOSCARDI, F. **Effect of soybean crop phenology on development, leaf consuption and oviposition of *Anticarsia gemmatalis* Hübner**. 1979.Thesis (Doctorate in Entomology) - University of Florida, UF, Gainesville.

MOSCARDI, F. Use of viruses to control the soybean caterpillar. In: ALVES, S.B. (Ed.) **Microbial control of insects**. Sâo Paulo: Manole, 1986. p.188-202.

MOSCARDI, F.; BARFIELD, C. S.; ALLEN, G. E. Consuption and development of velvetbean caterpillar as influenced by soybean phenology. **Environmental Entomology**, College Park, v. 10, n. 6, p. 880-884, 1981.

MOSCARDI, F.; CARVALHO, R.C.Z. Consumption and utilization of soybeans by *Anticarsia gemmatalis* Hüb. (Lepidoptera: Noctuidae) infected at different larval stages by its nuclear polyhedrosis virus. **Anais Sociedade Entomológica do Brasil**, Londrina, v.22, n.2, p.267-280, jun.1993.

MOSCARDI, F.; SOSA-GÓMEZ, D.R. A case study in biological control: soybean defoliating caterpillars in Brazil. In: INTERNATIONAL CROP SCIENCE CONGRESS, 1.,1992, Ames, Iowa. **Proceedings**...Ames: Crop Science Society of America, 1992. p.115-119.

MOSCARDI, F.; SOUZA, M. L. de. Baculovirus for pest control.

Biotechnology Science and Development. Londrina, n.24, p. 22-29, jan./feb.

2002.

NEVES, C.A., SERRĂO, J.E., GITIRANA, L.B. Ultrastructure of midgut endocrine cells in *Meipona quadrifasciata anthidioides* (Hymenoptera, Apidae). **Brazilian**

Journal of Biology. Sâo Carlos,v. 63, n.4, p.683-690, nov. 2003

OMYILAGHA, J.C.; GROTEWOLD, E. The biology and structural distribution of surface flavonoids. **Recent Research developments in plant science**, Columbus Ohio, v.2, p. 53-71, 2004.

O.REILLY, D.R.; MILLER, L.K.; LUCKOW, V. A. **Baculovirus expression vectors: a laboratory manual**. Salt Lake City, UT: W.H. Freeman, 1992.

PAINTER, R.H. **Insect resistance in crop plants.** New York, 1951.

PANIZZI, A. R.; et al. **Soybean insects in Brazil**. Londrina: EMBRAPA-CNPSo, 1977.

PANIZZI, A.R. Integrated management of soybean pests in Brazil. In: CROCOMO, W.B.

(Ed.). **Integrated pest management.** Sao Paulo: UNESP-FUNEP, 1990. p.293-321.

PATEL, N.; HAYES, T.K., COAST, G.M. Evidence for the hormonal function of a CRF-related diuretic peptide (*Locusta-DP*) in *Locusta migratoria*. **The Journal of Experimental Biology,** London, v.198, n. 3, p.793-804, mar. 1995.

PIUBELLI, G.C. **Bioactivity of soybean genotypes resistant to *A. gemmatalis* Hübner (Lepidoptera: Noctuidae) and interactions of their chemical substances with natural enemies**. 2004.Thesis (Doctorate in Biological Sciences - Entomology), Universidade Federal do Paranà, Curitiba, Paranâ. 152p.

PIUBELLI, G.C. Are chemical compounds important for soybean resistance to *Anticarsia gemmatalis* ? **Journal of Chemical Ecology**, California, v.31, n.7,p.1509- 1525, jul. 2005.

PRAÇA, L.B.; SILVA NETO, S.P.; MONNERAT, R.G. ***Anticarsia gemmatalis***

Hübner 1818 (Lepidoptera:Noctuidae) Biology, Sampling and Control Methods. 1. ed. Brasilia: Embrapa, 2006.

PRAÇA, L.B. et al. *Bacillus thuringiensis* strains effective against insects of the orders Lepidoptera,

Coleoptera and Diptera. **Pesquisa Agropecuâria Brasileira**, Brasilia, v.39, n.1, p. 11-16, jan. 2004.

PRATT, G.E. et al. Identification of allatostatin from adult *Diploptera punctata*. Biochemical and biophysical research communications, United States, v.163, n.3, p. 1243-1247, sep. 1989.

REID, J. C. **Larval development and consumption of soybean foliage by the velvetbean caterpillar, *Anticarsia gemmatalis* Hübner (Lepidoptera: Noctuidae) in the laboratory**.1975. Dissertation (Master's Degree in Entomology) - University of Florida, UF, Gainesville.

REZENDE, J.A.M.; ROSSETO, C.J.; MIRANDA, M.A.C. Behavior of paternal F1 soybean populations in relation to *Colaspis* sp. and *Diabrotica speciosa* (Germar, 1824). **Bragantia**, Campinas, v.39, n.,p. 15-20, 1980.

RICHARDS, O.W., DAVIES, R.G. **Imms' general textbook of entomology**. 10. ed. London: Chapman; Hall, 1994. v.1.

ROSSETO, C.J.; et al. Feeding preference of adults of *Diabrotica speciosa* (Germar) and *Cerotoma arcutata* (Oliv.) on soybean varieties. **Bragantia**, Campinas, v.40, n.1, p.179-183, jan. 1981.

SALVADOR, M.C. **Effect of soybean genotypes and flavonoids on the biology and midgut of *Anticarsia gemmatalis***. 2008. Dissertation (Master's Degree in Agricultural Entomology). Faculty of Agricultural and Veterinary Sciences - UNESP Jaboticabal Campus, Jaboticabal.

SCOOFS, L., VAN DEN BROECK, J., DE LOOF, A. The myotropic peptides of *Locusta migratoria*: structures, distribution, fuctions and receptors. **Insect biochemistry and molecular biology**. England, v.23, n.8, p.859-81, dec. 1993.

SETCHELL, K.D.R. Phytoestrogens: the biochemistry, physiology, and implications forhuman health of soy isoflavones. **The American Journal of Clinical Nutrition**, Cincinatti, v.68, n.1, p.1333S-1346S, jan. 1998.

SHILS. M.E. et al. **Treatise on modern nutrition in health and disease**. 1.ed. Tamboré: Manole Lda, 2003

SILVA, R. F. P. **Biological aspects and nutrition of *Anticarsia gemmatalis* Hübner (Lepidoptera - Noctuidae) in natural and artificial environments and the influence of temperature and photoperiod on its development**. 1981. Thesis (Doctorate in Entomology) - Escola Superior de Agricultura "Luiz de Queiroz"/USP, Piracicaba.

SILVA, R. F. P.; PARRA, J. R. P. Effect of photoperiod on the development of *Anticarsia gemmatalis* Hübner, 1818 (Lepidoptera: Noctuidae) under laboratory conditions. **Anais da Sociedade Entomológica do Brasil**, Londrina, v. 15, n. 2, p. 201-206, aug.1986.

SILVA, R.R. et al. Hypolipidemic effect of the flavonoids naringin and rutin. **Archivos Latinoamericanos de Nutrición**, Caracas, v.51, n.3, p. 258-264, Apr. 2001.

SILVA, R.R. et al. Effect of flavonoids on arachidonic acid metabolism.

Medicina, Ribeirao preto, v.35, p. 127-133, abr./jun. 2002.

SRIVASTAVA, R., Y.N. SHUKLA; S. KUMAR. Recent advances in the chemistry of insect antifeedants. **Journal of Medicinal and Aromatic Plant Sciences**, [S.l] v. 21, p. 59-76, 1999

SOSA-GOMES, D.R.Biological control of the soybean caterpillar with *Baculovirus anticarsia.* In: **PRÓ-GUAÌBA**: Subprogram of Management Systems and Control of Contamination by Pesticides. Porto Alegre: **EMATER-RS**, 1995, p.27-35.

SOSA-GÓMEZ, D.R. et al. **Manual de Identificaçao de Insetos e outros Invertebrados da cultura da soja.** Londrina: EMBRAPA-CNPSo, 2006.

SOSA-GÓMEZ, D.R. **Selectivity of agrochemicals for entomopathogenic fungi**. Available at <httpp://www.cnpso.embrapa.br/index>.

Accessed on: October 14, 2009.

STAMP, N.E.; SKROBOLA, K.L. Failure to avoid rutin diets results in altered food utilizatin and reduced growth rate of *Manduca sexta* larvae. **Entomologia Experimentalis et Applicata**, Netherlands, v. 68, n.2, p.127-142, aug. 1993.

STERN, V.M. et al., Improvements on self-propelled equipment for collecting, separating and tagging ,ass numbers of insects in the field. **Journal of Economic Entomology**, Baltimore, v.58 p.949-953, 1965.

TANADA, Y.; KAYA, H.K. **Insect pathology**. New York: Academic, 1993.

TARRAGÓ, M. F. S.; et al.Influence of ecological factors on the population fluctuation of soybean caterpillars, *Anticarsia gemmatalis* Hübner, and *Rachiplusia nu* (Guen.) in Santa Maria-RS. **Anais da Sociedade Entomológica do Brasil**, Londrina, v. 6, n. 2, p. 180-193, 1977.

TERRA,W.R.; FERREIRA, C. Insect digestive enzymes: properties, compartmentalization and fuction. **Comparative Biochemistry and Physiology**, Amsterdam, v.109B, p.1-62, n.1,jan. 1996.

TERRA, W. R. The origin and functions of the insect peritrophic membrane and peritrophic gel. **Archives of Insect Biochemistry and Physiology**, New York, v.47, n. 2, p.47-61, jun. 2001.

TOLEDO, T., et al., Effects of differents rutin dosages on chloride, calcium and phosphorum levels serum of rabbits. **Revista Chilena de Nutrición**, Santiago, v.30,n.3, p.263-270, dec. 2003.

VALADARES-INGLIS, M.C.; SOUZA, M.T.; SHILER, W. Genetic engineering of microorganisms that are agents of biological control In: MELO, L.S.; AZEVEDO, I. L. (Ed.). **Biological control**. Jaguariùna: Embrapa-CNPMA, 1998. v.1, p.208-217.

VALLE, G.E.; LOURENÇÂO. A.L. Resistance of soybean genotypes to *Bemisia tabaci* (Genn.) Biotype B (Hemiptera: Aleyrodidae). **Neotropical Entomology**, Vacaria, v.31, n.1, p.285-295, apr./jun. 2002.

VERNETTI, F.J. Soybean genetics: Qualitative characteristics. In: VERNETTI, F.J. (Ed.) **Soy:** genetics and improvement. Campinas: Cargill Foundation, 1983. p. 93124.

VOLKMAN, L.E.; KEDDIE, B.A. Nuclear polyhedrosis virus pathogenesis. **Seminars in Virology**, v.1, p.249-256, 1990.

VOLKMAN, L.E. Nucleopolyhedrovirus interactions with their insect hosts. **Advances in Virus Research**, Galveston,v.48, p.313-348, 1997.

WANG, P.; GRANADOS, R.R. Molecular structure of the peritrophic membrane (PM): identification of potential PM target sites for insect control. **Archives of Insect Biochemistry and Physiology**, New York, v.47, n.2, p.110-118, jun. 2001

WATSON, J. R. **The velvetbean caterpillar (*Anticarsia gemmatalis*)**. Florida: Report Florida Agricultural Experimental Station ,1915.

WATSON, J. R. Life-history of the velvetbean caterpillar *Anticarsia gemmatalis* Hübner. **Journal of Economic Entomology**, New York, v. 9, p. 521-528, 1916.

WILLE, J. E. **Entomologia Agricola del Peru**. Lima: Estacion Experimental Agricola de la Molina, 1943.

WOODHEAD, A.T.; et al. Primary structure of four allatostatins: neuropeptide inhibitors of juvenile hormone synthesis. **Proceedings of the National Academy of Sciences of the United States of America**, Iowa City, v.86, n. 15, p.5997-6001, Aug. 1989.

3 ARTICLE A: POST-INGESTIVE EFFECTS OF THE FLAVONOIDS RUTIN AND GENISTIN ON THE MIDDLE INTESTINE OF *A. GEMMATALIS* LARVAE SUSCCEPTABLE AND RESISTANT TO THE NUCLEOPOLIEDROVÌRUS *AG* MNPV

SUMMARY

The soybean caterpillar *Anticarsia gemmatalis* (Lepidoptera:Noctuidae) is one of the main pests of the soybean crop, and when not controlled can cause up to 100% crop losses. It is controlled by spraying a bioinsecticide based on the multiple nucleopolyhedrovirus (*AgMNPV*) and by using plant resistance with cultivars that have a complex of chemical substances, mainly flavonoids that make the plants toxic to insects. As the midgut is the main route for insecticides and chemical substances from resistant plants, it is necessary to study the effect of flavonoids on this organ. The aim of this study was to verify the morphological changes in the midgut *of A. gemmatalis* larvae susceptible and resistant to *AgMNPV* after ingesting leaves of soybean genotypes containing rutin (R) and genistein (G) at different concentrations. Both larval populations came from Embrapa-CNPSoja/Londrina-PR and were treated with insect-resistant "in natura" soybean genotypes: BRS 257 (control); BR 16, Dowling, PI 229358, IAC 100 and PI 227687. The midguts of the 4th instar larvae were fixed in Karnovsky, processed and analyzed under a light microscope. All the treatments caused changes in the epithelial cells of the midgut. The columnar cells were the most altered, showing great vacuolization and cytoplasmic voids and the release of protrusions with cellular content in both larval populations. The goblet cells were little altered, while the regenerative cells of the susceptible larvae in the Dowling treatment and the resistant larvae in the IAC 100 treatment showed intense activity at the base of the epithelium. The Dowling, IAC 100 and PI 227687 treatments caused specific changes in the muscular layers of the LS, and the PI 229358, IAC and PI 227687 treatments in the muscular layer of the resistant larvae. Among the treatments, the IAC 100 genotype caused the greatest changes in the IM of resistant larvae and the PI 227687 genotype caused the greatest changes in susceptible and resistant larvae.

Keywords: Soybean caterpillar. Resistance. Protrusions. Midgut, Genotypes.

1 INTRODUCTION

Brazil is responsible for a large part of the world's grain production and soybeans (*Glycine max*) are considered Brazil's main export crop (EMBRAPA, 2005; CONAB, 2009). It is the plant species of greatest economic importance, making the country the second largest producer and exporter of soybean meal and grain in the world, reaching a production of 62 million tons in the last harvest (CONAB, 2009).

Pest insects are important factors in reducing productivity in different cultivated plant species. In soybeans, the larvae *of Anticarsia gemmatalis* Hübner (Lepidoptera: Noctuidae), known as the "soybean caterpillar", are one of the main defoliating pests and in Brazil, they are distributed in practically all growing areas. Under normal conditions, this pest occurs in crops from November in the northern regions of Paranà, and from December to January throughout the south of the country, causing defoliation that can reach 100% (HOFFMANN-CAMPO et al., 2000; SOSA-GÓMEZ et al., 2006; DEGRANDE; VIVAN, 2006).

Pest infestations in crops are generally controlled through the use of pesticides, but their indiscriminate use is ecologically harmful and can lead to insect resistance to the products used, outbreaks of secondary pests, adverse effects on beneficial organisms, undesirable residues in food and direct damage to product applicators (HOFFMANN-CAMPO, 1995). In order to reduce the use of chemical control, alternative techniques have been used to combat this pest, such as the use of biological insecticides and the use of resistant plants (HOFFMANN-CAMPO et al., 2000; 2001; PIUBELLI et al., 2005).

The bioinsecticide based on the virus *AgMNPV* (*Anticarsia gemmatalis* multiple nucleopolyhedrovirus), specific for this pest, has been used as a biological control since the 1980s (MOSCARDI; CARVALHO, 1993; MOSCARDI, 1999; ABOT et al., 1996). Infection with *AgMNPV* begins when the caterpillars ingest the viral polyhedra together from the soybean leaves; the virions are released into the alkaline lumen of the midgut, pass through the peritrophic membrane and fuse with the membrane of the epithelial cells, after recognizing specific receptors, beginning the replication cycle (FLIPSEN et al., 1995; BARRET et al., 1998).

The use of plant cultivars with resistance to pests is more recent and is considered to be the ideal method of control, since populations can be reduced below the level of economic damage, without environmental pollution (LARA, 1991), in addition to not requiring the adoption of sophisticated technology by farmers and not presenting additional costs (EMBRAPA, 2005).

Soybean resistance to insects is a tactic widely used in pest control programs. The plant's defense mechanisms include a series of morphological characteristics and also a complex of chemical substances that can make it repellent, toxic or otherwise unsuitable for insect pests (PIUBELLI, 2004). In soybeans, the most likely defense substances are flavonoids (HOFFMANN-CAMPO, 1995), which are generally more abundant in pest-resistant genotypes. Flavonoids are a group of natural substances that have very diverse biological activities. One of the major actions of flavonoids is related to the formation of fatty acids by the action of phospholipase A2, responsible for the hydrolysis of phospholipids present in cell membranes, with the release of arachidonic acid (KUBO; HANKE, 1986; LARA, 1991; SILVA et al., 2002; ONYILAGHA; GROTEWOLD, 2004; LOPES et al., 2005). Thus, flavonoids can interact and penetrate the lipid bilayers of membranes, altering the barriers and permeability of cell membranes (LIMA et al., 1999; SILVA et al., 2001). Other functions could be related to acting on the expression of membrane receptors, as antibacterial and antiviral agents and inhibitors of proteolytic enzymes (TOLEDO et al., 2003).

The phenolic compounds present in plants have different effects on herbivores. The activation of these compounds occurs by oxidation and can lead to the precipitation of proteins and inhibition of digestion can occur due to the formation of hydrogen bridges, or by the formation of covalent bonds with proteins and digestive enzymes (APPEL, 1993). The hydroxyl radicals formed during the oxidation of phenolic compounds have a toxic action and are responsible for disrupting membrane integrity and metabolic disorders in the intestinal epithelium (APPEL, 1993). In caterpillars, the covalent bond between chlorogenic acid and monomeric proteins inhibits the absorption of amino acids.

(FELTON et al., 1989; FELTON; DUFFEY, 1991).

Seeking to understand the effects of flavonoids on insects, Hoffmann-Campo et al. (1994) carried out food preference tests on *A. gemmatalis* and noted that the soybean genotypes BR82-12547, IAC74-2832, PI 227687, PI 229358, PI 274454 were the most rejected by the caterpillar. Later studies identified the presence of seven flavonoids in PI 227687, one of the most widely used in insect resistance programs, which could indicate greater toxicity of this cultivar and explain the insect's rejection. Among the flavonoids, rutin (quercitin 3-O-rutinoside) is known to play a very important role in plant defence against lepidopterans (HOFFMANN-CAMPO, 1995; HOFFMANN-CAMPO et al., 2001). This substance has also been identified in extracts from the leaves of PI 227687, PI 274454 (PIUBELLI et al., 2005). Previous studies of these substances on defoliator insects have shown antibiotic and/or non-consumptive effects on Lepidoptera larvae such as *Manduca sexta* (L.) (STAMP; SKROBOLA, 1993), *Heliothis virescens* (F.) (HOFFMANN-CAMPO, 1995) and *Trichoplusia ni* (Hübner) (HOFFMANN-CAMPO et al., 2001).

To act effectively and efficiently, both the virus (*Ag* MNPV) and the chemical substances must be ingested by the caterpillars, and after ingestion, the virus and the chemical substances will act directly in the digestive tract, which in insects is considered an efficient physical-chemical barrier against the action of microorganisms that may be ingested together with the food (TERRA; FERREIRA, 1991; CHAPMAN, 1998). In insects, the main reactions in the digestion and absorption of food, as well as chemical and biological plant protection products, take place in the midgut (TERRA; FERREIRA, 1991; CHAPMAN, 1998).

There are few studies on the nutritional and post-ingestive effects of flavonoids on the soybean caterpillar. According to Gazzoni et al (1997) in *A. gemmatalis,* rutin and genistein showed effects characteristic of antibiosis, lengthening the larval cycle and reducing survival. Piubelli et al (2006) found that rutin increased the mortality of *A. gemmatalis* larvae resistant to the *Ag* virus MNPV compared to susceptible larvae. Recently, Salvador (2008) found morphological changes in the epithelium of the midgut (MI) in *A. gemmatalis* as a result of ingesting rutin and genistein, and leaves of soybean genotypes, with and without insect resistance characteristics, which have these substances in their chemical composition. The substances were added to artificial diets in different concentrations, and the effects of insect-resistant soybean genotypes (PI 227687, IAC-100, PI 274454 and Dowling), which have rutin and genistein in their composition, and (BR-16) considered to be susceptible, without rutin in its composition were evaluated. For the author, the best effect was detected when rutin and genistein acted together, leading to epithelial disruption of the IM. According to Salvador (2008), this is the first study to look at the mode of action of these compounds in the IM of *A. gemmatalis*, which is why morphological, histochemical and ultrastructural studies will be necessary to better understand the action of flavonoids in the intestinal tract of the soybean caterpillar.

In *A. gemmatalis* larvae, the gut is formed by a pseudostratified columnar epithelium with at least four distinct cell types: columnar, goblet, regenerative and endocrine cells (LEVY et al., 2004, 2005). The epithelium is lined by an acellular structure called the peritrophic membrane (PM), which separates the epithelium from the intestinal contents, protecting it from abrasion, acting as a permeable barrier to digestive enzymes and digestive products (EISEMANN; BINNINGTON, 1994; TERRA, 2001), as well as forming a barrier that hinders or prevents microorganisms from entering the hemocele (WANG; GRANADOS, 2001). Below the epithelial layer, supporting all this architecture and separating the digestive tract from the rest of the hemocele, is a delicate basement membrane of conjunctival origin (only visible under a transmission electron microscope) and the muscular layer, which is made up of two types of muscle: longitudinal and circular (LEHANE 2006; GALLO et al., 2002; LEVY et al., 2004).

When the entire structure of the midgut is attacked by pathogens or chemical substances, it undergoes changes in its morphology, physiology and metabolism, resulting in weaknesses that will not allow the caterpillar to complete its development cycle, and sometimes even cause its death (FUGI et al., 2005; SALVADOR, 2008).

Although the biology of the soybean caterpillar has been extensively studied, there is still little information on the midgut of this insect in the specific literature, with the exception of Levy et al. (2004), who described the digestive system of AgMNPV-susceptible and -resistant larvae fed diets free of or inoculated with the AgMNPV virus. However, there are no studies on the changes in the midgut of *A. gemmatalis* larvae susceptible and resistant to *AgMNPV*, concomitant with the ingestion of chemical substances (flavonoids) found in different soybean genotypes with insect resistance characteristics.

As the midgut of insects is the main route of access and action of these substances, there is a need to study the alterations caused by them. The aim is to correlate the post-ingestive effects of the flavonoid rutin and the isoflavonoid genistein, both alone and in combination, in the midgut of *A. gemmatalis* larvae and to verify the possible effects of these substances in breaking down the resistance of these larvae to *AgMNPV*.

2 MATERIAL AND METHODS

The following soybean genotypes were used: BRS 257, BR 16, Dowling, PI 229358, IAC 100 and PI 227687. The genealogy of the genotypes and the concentrations of flavonoids rutin and genistein present are described in Table 2.1.

Table 2.1 - Genealogy of the soybean genotypes used and the concentrations of flavonoids rutin and genistein in mg/g x 10 .$^{-2}$

Treatments	Genealogy	Concentration (mg/g)	
Genotypes		Rutin	Genistina
1. BRS 257	BR93-32109 x BR94-23396	0,0	0,0
2. BR 16	D69-1310-M58 (D49-772 x D49-2491) x 'Davis'	0,0	0,16
3. Dowling	Semmes x PI 200492	0,0	4,29
4. PI 229358	Tokyo, Japan	2,12	1,36
5. IAC 100	'IAC 12' x IAC 78-2318	9,72	1,42
6. PI 227687	Okinawa, Japan	36,82	1,22

Source: Valle and Lourençâo, (2002)

The control genotype (BRS 257), the genotype (BR 16) with insect susceptibility characteristics (SALVADOR, 2008) and the insect-resistant genotypes (Dowling, PI 229358, IAC 100 and PI 227687) were sown in a shaded greenhouse with daylight at Embrapa Soja Londrina-PR, in pots containing soil and drip irrigation with a capacity of five liters, with five seeds per pot, in a total of seven pots/treatment No top dressing or application of phytosanitary products was carried out. The leaves were collected, taken to the laboratory and washed in a 5% sodium hypochlorite solution; rinsed in distilled water; dried on filter paper and placed in a gerbox with *A. gemmatalis* eggs.

The eggs of *A. gemmatalis* susceptible and resistant to AgMNPV were supplied by the Entomology Laboratory of EMBRAPA Soja, Londrina/PR. To obtain *A. gemmatalis* eggs susceptible to *AgMNPV*, the larvae were kept on an artificial diet (HOFFMANN-CAMPO et al., 1985) at a controlled temperature (25-27°C), photoperiod (14h light/10h dark) and 80% relative humidity. After reaching the pupal stage, they were separated into males and females and placed in cages with 60 to 70 pairs to emerge and mate. The cages were lined with off-set sulphite paper with a waterline for oviposition. The papers with the eggs were immersed in water; the eggs were carefully peeled off, sieved through fine mesh sieves and placed on pieces of paper (3x4 cm), with approximately 200 pieces. *AgMNPV-resistant A. gemmatalis* eggs were obtained through the same process described above, modifying the diet offered to the larvae, which contained a concentration of *Baculovirus* maintained under selection pressure (Lethal Concentration - CL_8 O in each generation), according to the procedures of Abot et al. (1996).

The papers containing the eggs were fixed to paper towels and placed on the lid of the gerbox, in a total of four gerboxes for each treatment (genotype), totaling around 800 eggs/treatment. The gerboxes had filter paper at their base to make it easier to keep the larvae and clean them. In addition, 2 leaves of the respective genotypes were placed on the filter paper. To keep the leaves moist and ensure tender leaves, portions of moistened cotton were added to each petiole.

After hatching, the larvae were monitored daily until the second instar, when they were replanted in six gerboxes containing around 25 larvae per gerbox, to facilitate rearing. The soya leaves were changed every 12 hours so that the larvae didn't run out of food.

After the 4th instar (10 to 12 days of post-hatch development), *A. gemmatalis* larvae susceptible and resistant to AgMNPV were anesthetized and immobilized by cooling (5 minutes). *gemmatalis larvae* susceptible and resistant to AgMNPV were anaesthetized and immobilized by cooling (5 minutes) and dissected under a stereoscope with saline solution for insects (1.80 g NaCl; 1.88 g KCl; 0.16 CaCl; 0.004 g $NaHCO_3$ and distilled water-q.s.p 100 mL). The larvae were pinned at the ends by entomological pins, in a Petri dish filled with paraplast and with the help of ophthalmic scissors, a logitudinal incision was made in the ventral region, from the first thoracic segment to the penultimate abdominal segment, with the cuticle folded back laterally. The tubes, 30 in number per treatment, were collected, fixed in Karnovsky's fixative solution (2.5% glutaraldehyde + 4.0% paraformaldehyde in 0.1M phosphate buffer and pH 7.2) for 6 hours and processed according to the protocol of Cerri and Cerri-Sasso (2003).

After fixation, the material was washed in sodium phosphate buffer solution (0.1M and pH 7.2) for 5 minutes; dehydrated in 90% ethyl alcohol (20 minutes) and 100% ethyl alcohol (20 minutes); pre-infiltrated in a solution of basic resin (hydroxyethyl methacrylate) + 100% alcohol (1. 1) for 4 hours at room temperature:1) for 4 hours at room temperature; infiltrated in pure resin (Leica Historesin soaking kit®) at room temperature for 24 hours; embedded in appropriate polyethylene molds containing a solution of basic resin (hydroxyethyl methacrylate) + activator (dibenzoylperoxide) + hardener (dimethylsulfoxide) and kept at room temperature until the resin polymerized. The embedded material was cut to a thickness of 4µm with a tungsten knife on a LEICA RM 2265 automatic rotary microtome and placed on glass slides previously cleaned with ether alcohol.

The sections were hydrated (5 minutes) in distilled H2O, stained with Harris' Hematoxylin (40 minutes); washed in running water (10 minutes), stained with aqueous Eosin (30 minutes), washed in distilled water (5 minutes); dehydrated in 95% and 100% alcohol; diaphanized in xylene (30 minutes); mounted in Canada balsam and after drying they were analyzed in the light *photomicroscope* (Axiophot by Zeiss) of the Histology Department at UEL.

3 RESULTS

The *A. gemmatalis* larvae, susceptible (LS) and resistant (LR) to the nucleopolyhedrovirus *Ag* MNPV, fed on the leaves of the different soybean genotypes, showed similar midgut architecture, but with specific variations in many of its components.

The morphological description of the IM of the LS and LR fed with leaves of the BRS 257 genotype (R=0; G=0) was the parameter of comparison for the different genotypes, since it did not have a concentration of flavonoids (rutin and genistein) in its chemical composition, and was therefore the experiment's control.

The IM of the LS and LR treated with the BRS 257 genotype consisted, from the lumen to the periphery, of the peritrophic membrane, epithelial layer and muscular layer (Figs. 4.1 A and 4.2 A).

The peritrophic membrane was a thin acellular structure, very visible and preserved throughout the midgut, dividing the intestinal lumen into endoperitrophic and ectoperitrophic spaces. The endoperitrophic space contained a large amount of food and the ectoperitrophic space was small and very close to the epithelium.

The epithelium of the midgut in both larval types was of the prismatic pseudostratified type with a striated border, formed by three distinct cell types: columnar cells (Co); goblet cells (Ca) and regenerative cells (Cr) (Figs. 4.1 A , 4.2 A).

The columnar cells were tall, prismatic and predominated along the length of the midgut. The cytoplasm was basophilic, vacuolized with an elongated nucleus, generally located in the upper third of the cell, between the central and apical regions. These cells had distinct cytoplasmic regions which allowed them to be differentiated into the apical region, with vacuolization and cytoplasmic expansions forming the striated border; the middle region with less vacuolization than the apical portion and the basal region with a more homogeneous appearance and few vacuoles (Figs. 4.1 A and 4.2 A).

The goblet cells had a large goblet-shaped cavity with a small amount of acidophilic cytoplasm and a flat, basal nucleus (Figs. 4.1 A and B, 4.2 A).

The regenerative cells were found isolated or in small groups at the base of the epithelium and were rounded or oval in shape, with basophilic cytoplasm and a central nucleus (Figs. 4.1 A).

Below the epithelial layer, the muscular layer was observed, consisting internally of bundles of transverse muscle fibers from the circular muscle and externally of longitudinally arranged muscle bundles from the longitudinal muscle (Fig 4.1 A).

3.1 Larvae Susceptible to MNPV *Ag*

The degree of alteration of the midgut structures of the susceptible larvae and the respective concentrations in mg/g of the flavonoids rutin and genistein present in the soybean genotypes are described in Table 3.2.

Table 3.2 - Evaluation of the changes caused by rutin and genistein in the midgut of larvae susceptible (LS) to the nucleopolyhedrovirus (*Ag* MNPV) and the concentrations of flavonoids in mg/g X 10 .$^{-2}$

	Genotypes					
Structures	**BRS 257** **R: 0** **G:0**	**BR 16** **R: 0** **G:0,16**	**Dowling** **R: 0** **G:4,29**	**PI229358** **R: 2,12** **G: 1,36**	**IAC 100** **R:9,72** **G: 1,42**	**PI227687** **R: 36,82** **G:1,22**
Peritrophic membrane	-	-	-	-	+++	+++
Grooved edge	-	-	+++	+++	+	+

Cells	Apical portion	-	+++	+++	+++	+++	++
Columnar	Average portion	-	+++	+++	+++	+	+++
	Basal Portion	-	+	++	+	+++	+++
	Protrusions	-	+++	+++	+++	+++	+++
Goblet cells		-	++	+++	++	++	+++
Regenerative cells		-	++	+++	-	-	+++
Muscle Layer		-	-	+	+++	+++	-

Legend: (-) no change; (+) slight change; (++) moderate change; (+++) strong change

The IM of the LS treated with the BR 16 genotype (G = 0.16×10^{-2} mg/g; D_1 G , considering dose 1 or initial dose of genistein) compared to the BRS 257 genotype, showed strong alterations in the epithelial layer, especially in the columnar cells, which showed cytoplasmic protrusions being released in large quantities towards the intestinal lumen. These protrusions had cytoplasmic content, as well as some cell nuclei. These cells showed alterations in the different cytoplasmic portions, with the apical and middle portions showing intense vacuolization, with the formation of large voids (Figs. 4.1B) while the basal portion showed few vacuoles.

The goblet cells appeared to be very numerous, sometimes taller and narrower with the release of product into the lumen and sometimes bulky (Fig. 4.1 B); the regenerative cells were also quite numerous.

In the treatment with the Dowling genotype (G = 4.29×10^{-2} mg/g; D_{2G} = dose of genistein 27X D_1), the epithelium showed strong alterations in the columnar cells, including impairment of the striated edge, making it difficult to recognize this structure. The apical and middle portions of the columnar cells showed a great deal of vacuolization. In some parts of the epithelium, columnar cells with nuclei were detached towards the lumen (Fig. 4.1C). The goblet cells showed strong changes, apparently more voluminous and with the release of cavity contents into the intestinal lumen (Fig. 4.1C), while the regenerative cells were present in large numbers forming groups of 3 to 5 cells (Fig. 4.1 C). The muscle layer showed little change and was apparently thicker than the muscle layer of the control.

The peritrophic membrane showed no changes in morphology. It was observed that the endoperitrophic space contained a large amount of food, but the ectoperitrophic space was well away from the epithelium.

The midgut of larvae treated with the PI 229358 genotype (G = 1.36×10^{-2} mg/g; D_{3G} = dose of genistein 8.5 X D_1) and (R = 2.12×10^{-2} mg/g; D_{1R} = initial dose of rutin) showed changes in the epithelium and muscle layer (Fig. 4.1D).

Among the cell types of the epithelium, the columnar cells were the most affected, with thickening of the striated border, intense vacuolization of the apical and middle portions and numerous protrusions being released into the intestinal lumen (Figs. 4.1 D). Fragments of cytoplasm were observed in these protrusions, with the presence or absence of a nucleus, and in some cases they were covered by the striated border (Fig. 4.1D).

The goblet cells and regenerative cells did not show any morphological changes, but the goblet cells were observed in large numbers throughout the epithelium of the midgut (Fig. 4.1 D).

The muscle layer showed changes, with internal fragmentation (Fig. 4.1D).

There was no apparent damage to the peritrophic membrane, but the endoperitrophic space was very small, with little lumen and no food, while the ectoperitrophic space was very wide and far from the epithelium (Fig. 4.1D).

In the IAC 100 treatment (G = 1.42 x 10^{-2} mg/g; D $_{4G}$ = dose of genistein 8.8 X $_{D1}$) and (R = 9.72 x 10^{-2} mg/g; D $_{2R}$ = 4.6 X initial dose of rutin) there were changes in all the structures that make up the wall of the midgut.

In the columnar cells, a large number of vacuoles were observed in the apical and basal portions (Figs. 4.1 E), as well as the discarding of whole cells into the lumen (Fig. 4.1 E) with a large number of cytoplasmic protrusions inside and portions of cytoplasm, cell nuclei were observed. The goblet cells were more rigid and showed an expanded acidophilic border inside the cell apex (Fig. 4.1 E).

It was not possible to observe the presence of the peritrophic membrane or the spaces delimited by it, while the muscular layer was evidenced only by small fragments.

Treatment with the PI 227687 genotype (G = 1.22 x 10^{-2} mg/g; D $_{5G}$ = dose of genistein 7.6 X $_{D1}$) and (R = 2.12 x 10^{-2} mg/g; D $_{3R}$ = 17.4 X initial dose of rutin) caused the greatest changes in the epithelial layer. The columnar cells showed great vacuolization in all their cytoplasmic portions and sometimes the vacuoles formed large cellular voids. In addition, these cells showed a large number of protrusions containing cytoplasm and cell nuclei. The striated border was apparently thicker and easier to see (Fig. 4.1 F).

The goblet cells showed alterations mainly in their cytoplasm (Fig. 4.1 F) and the regenerative cells were found in greater numbers.

It was not possible to visualize the peritrophic membrane, while the muscular layer showed no alterations.

3.2 LARVAE RESISTANT TO AGMNPV

The degree of alterations in the midgut structures of the resistant larvae and the respective concentrations expressed in (mg/g) of the flavonoids rutin and genistein present in the soybean genotypes are described in Table 3.3.

Table 3.3 - Evaluation of the changes caused by rutin and genistein in the midgut of larvae resistant (LR) to the nucleopolyhedrovirus (*Ag* MNPV) and the concentrations of flavonoids in mg/g x 10^{-2}.

	Genotypes					
	BRS 257	**BR 16**	**Dowling**	**PI229358**	**IAC 100**	**PI227687**
Structures	**R: 0**	**R: 0**	**A: 0 G:4,29**	**R: 2,12**	**R:9,72**	**R: 36,82**
	G:0	**G:1,6**		**G: 1,36**	**G: 1,42**	**G:1,22**

Peritrophic membrane		-	-	-	++	-	+++
Cells	Grooved edge	-	-	+++	+++	+++	+
Columnar	Apical portion	-	+	+++	++	+++	+++
	Average portion	-	+++	+	+++	+++	+++
	Basal Portion	-	++	+++	+	+	+
	Protrusions	-	+++	+++	+++	+++	+++
Goblet cells		-	-	+++	+++	++	+++
Regenerative cells		-	+	-	-	+++	+
Muscle Layer		-	-	-	+++	+++	+++

Legend: (-) no change; (+) slight change; (++) moderate change; (+++) strong change.

The midgut of resistant larvae treated with the BR 16 genotype showed alterations only in the epithelial layer and only in the columnar cells, which showed vacuolization in the middle and basal portions, the former being the most vacuolated (Fig. 4.2 B). A large number of protrusions with cytoplasmic and nuclear content were visualized.

In general, treatment with the Dowling genotype led to changes in the epithelial layer, with the columnar cells and the striated border being preserved and very thick, and with the release of cytoplasmic protrusions towards the intestinal lumen (Figs. 4.2C). The calyces were apparently in greater quantity when compared to the larvae treated with the control genotype BRS 257 (Fig. 4.2 C).

In the treatment with the PI 229358 genotype, great disorganization was observed in most of the structures analyzed. Among the epithelial cells, the regenerative cells apparently showed no morphological changes (Fig. 4.2D). The other cells (columnar and goblet cells) showed significant alterations. The columnar cells showed alterations with the presence of vacuoles in all three portions, with more vacuolization in the basal and apical portions. The striated border was visible, but disorganized, with a large number of protrusions containing cytoplasm and nuclei (Fig. 4.2D).

The peritrophic membrane and muscular layer were observed, fragmented, throughout the entire length of the intestine (Fig. 4.2D).

In LR treated with the IAC 100 genotype, changes were seen in the three cell types that make up the epithelium. What was striking were the goblet cells, which sometimes appeared larger, with visualization of the release of contents from the calyx into the intestinal lumen (Figs. 4.2E). The regenerative cells were observed individually or in groups of up to 6 cells (Fig. 4.2E) and apparently increased in size. The peritrophic membrane showed no apparent changes, while the muscle layer was very thick (Fig. 4.2E).

In the PI 227687 treatment, alterations were identified in all the layers that make up the IM, being more intense in the epithelial layer, where the columnar cells were very altered. Vacuolization was present in all portions, with the basal portion being less vacuolized than the apical and middle portions. The latter showed large vacuoles and cell voids, especially in the middle portion (Fig. 4.2 F). The striated border was thick and its microvilli were clearly visible. The protrusions were found in large numbers detached from the intestinal lumen and containing cytoplasm and cell nuclei inside. The goblet cells had an opening towards the lumen and with the release of their contents, in some situations this goblet was more voluminous, and the regenerative cells were numerous (Fig. 4.2 F). It was not possible to see the peritrophic membrane, while the muscle layer was fragmented and thin (Figs. 4.2 F).

4 PHOTOGRAPHIC DOCUMENTATION

A: BRS 257 treatment (control): General morphology of MI. Columnar cells (Co) with brush border (Be); Goblet cells (Ca); Regenerative cells (Cr); Muscle layer composed of longitudinal muscle (Ml) and circular muscle (Mc). Peritrophic membrane (Mp). Bar = 10μm.

B: Treatment BR 16: Columnar cells (Co) with cytoplasmic protrusions (P) being released into the intestinal lumen, presence of cytoplasmic vacuoles (V) of various sizes; Goblet cells (Ca) apparently in greater quantity and higher than the control; Regenerative cells (Cr); Muscle layer (Cm) preserved and peritrophic membrane (Mp).Bars = 10μm

C: Dowling treatment: Columnar cells (Co) with a spongy appearance and detached towards the lumen; presence of cytoplasmic vesicles (v); goblet cells (Ca); and group of regenerative cells (Cr); fragmented muscle layer (Cm). Bar = 100μm.

D (PI 229358 treatment); E (IAC 100 treatment) and F: (PI 227687 treatment): Columnar cells (Co), vacuolized (V), with cytoplasmic protrusions (P), with the presence of a nucleus (n) and a brush border (Be); Goblet cells (Ca) in large quantities; Regenerative cells (Cr); Muscle layer (Cm). Note the presence of a peritrophic membrane (Mp) in the PI 229358 treatment; the presence of columnar cells discarded whole into the lumen (→) in the IAC 100 treatment; goblet cells (Ca) with cytoplasmic alteration and the presence of secretion (★) in the PI 227687 treatment; and a very fragmented muscle layer (Cm) in the PI 229358 and IAC 100 treatments. Bars (D and F) = 10μm; (E) = 100 μm.

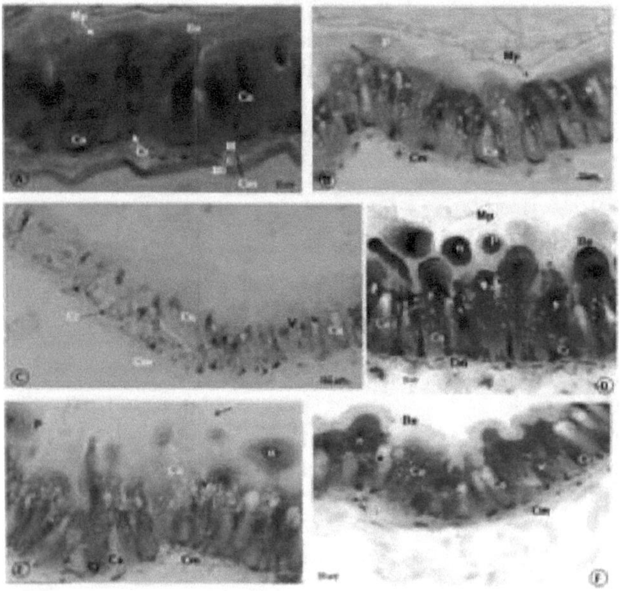

Figure 4.1 - Photomicrographs of the midgut *of A. gemmatalis* larvae susceptible to the *Ag* MNPV. Hematoxylin and Eosin (HE) staining.

A: BRS 257 treatment - General morphology of MI. Columnar cells (Co) with cytoplasmic vacuoles (V) and brush border (Be); goblet cells (Ca); regenerative cells (Cr); muscle layer (Cm). Peritrophic membrane (Mp), ectoperitrophic space (Ec) and endoperitrophic space (Ed). Bar = 10μm.

B (BR 16 treatment); C (Dowling treatment) and D (PI 229358 treatment) - Columnar cells (Co) with many vacuoles (v); covered with a thick brush border (Be); cytoplasmic protrusions (P) and with a cell nucleus (n); goblet cells (Ca); regenerative cells (Cr); and muscle layer (Cm). Note the presence of peritrophic membrane fragments (Mp) in treatment PI 229358. Bar = 10μm.

E (IAC 100 treatment) and F (PI 227687 treatment) - Columnar cells (Co) with cytoplasmic vacuoles (V); covered with brush border (Be) and release of cytoplasmic protrusions (P) with cell nucleus (n); voluminous goblet cells (Ca); thick muscle layer (Cm). Note the presence of regenerative cell nests (NCr) and the peritrophic membrane (Mp) in the IAC 100 treatment. Bars = 10μm.

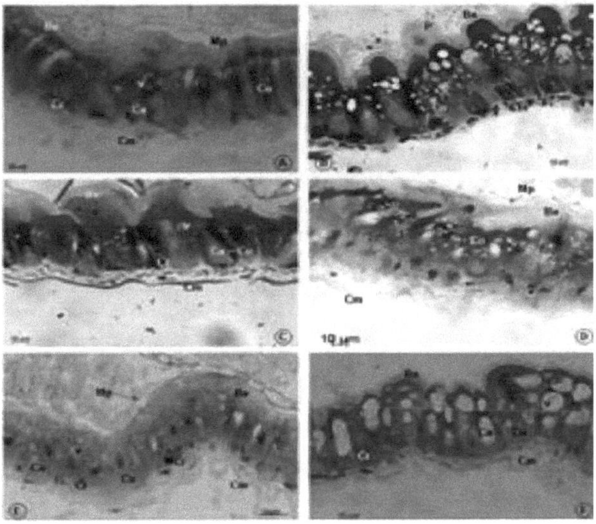

Figure 4.2 - Photomicrographs of the midgut *of A. gemmatalis* larvae resistant to *AgMNPV*. Hematoxylin and Eosin (HE) staining

5 DISCUSSION

In the treatment with the control genotype BRS 257 (G=0 and R=0), the midgut *of A. gemmatalis* larvae susceptible and resistant to AgMNPV showed a morphology similar to those described in the literature. The midgut was formed by a prismatic pseudostratified epithelium, with a striated border and composed of three distinct cell types: columnar, goblet and regenerative cells. This epithelial layer was covered by an acellular structure, the peritrophic membrane, which separated the epithelium from the intestinal contents and divided the lumen into two spaces, the ectoperitrophic and the endoperitrophic. Below the epithelial layer, supporting all this architecture and separating the digestive tube from the rest of the hemocele, was the muscular layer, made up of two types of muscle: longitudinal and circular. This organ architecture is the same as that described for most lepidopterans by various authors (EISEMANN; BINNINGTON, 1994; LEHANE; BILLINGSLEY, 1996; GALLO et al., 2002; LEVY 2000; 2004; LEVY et al., 2005; LEHANE 2006; SALVADOR 2008; SOUSA et al., 2009).

Our results showed that the two larval populations, both susceptible and resistant to *AgMNPV*, fed on the leaves of the genotypes BR 16, Dowling, PI229358, IAC 100 and PI 227687, which contained the flavonoid rutin and\or the isoflavonoid genistein, showed morphological changes in the midgut, varying in intensity according to the concentration of these compounds. These changes, although occasional, were quite observable and very distinct, ranging from the presence of small vacuoles between the columnar cells; an increase in cytoplasmic protrusions containing cellular content inside them and even the total detachment of entire cells into the intestinal lumen.

It is known that when pathogens or chemical substances attack the entire structure of the midgut, its morphology, physiology and metabolism are altered, resulting in weaknesses that will not allow the caterpillar to complete its development cycle, and sometimes even cause its death (FUGI et al., 2005; SALVADOR, 2008).

The greatest alterations were observed in the epithelial layer of the IM of both LS and LR and among the cell types that make up the epithelium, the one that suffered the greatest alteration was the columnar cell. These changes were observed in the cell protrusions and in the apical and middle regions of these cells in the two larval types treated with all the genotypes (BR16, Dowling, PI229358, IAC 100 and PI 227687). Even the BR16 genotype, which was considered by Piubelli et al (2005) to be susceptible to caterpillars because it did not contain rutin and had a low concentration of genistein, showed the ability to alter the epithelium.

The two types of larvae treated with the different genotypes showed intense cytoplasmic vacuolization in the columnar cells, the presence of "cytoplasmic voids" and the release of cytoplasmic protrusions without and with cell content, mainly nuclei, towards the ectoperitrophic space. These facts were similar to those described by Rharrabe et al. (2007) who, studying the action of beta-carboline allelochemicals (alkaloids) from *Peganum harmala* (Syrian rue), observed cytotoxic effects on the epithelial cells of the midgut *of Plodia interpunctella* (Lepidoptera: Pyralidae) and that these cells showed cytoplasmic voids, intense vacuolization and the release of cytoplasmic protrusions.

Salvador (2008) checked the effects of the flavonoid rutin and the isoflavonoid genistein on fourth instar *A. gemmatalis* larvae fed an artificial diet and the soybean genotypes BR16, PI 227687, IAC 100, PI 274454 and Dowling. Although the author used a higher concentration of the two chemical components than the concentrations used in our work (up to 10 X), she reported changes similar to those observed by us in the columnar cells of larvae susceptible to *Ag* MNPV.

Studies carried out by Arruda et al. (2008) on *Aedes aegypti* (Diptera) larvae treated with an extract of *Magonia pubescens*, popularly known as tingui, a tree found in cerrado areas, showed that this extract caused major changes in the midgut of the larvae tested, with intense vacuolization and cytoplasmic voids in the cells. Silva et al. (2004) described the larvicidal effects of tannins isolated from tingui on *A.* aegypti, while studies carried out by Rey; Pautou; Meyran (1999) using histological sections showed the toxic effect of tannic acid, a hydrolysable natural tannin, on the epithelium of the midgut of dipteran larvae. Other authors such as Pautou et al. (2000), after evaluating the toxic effects of tannins on the fauna associated with culicids, suggested that plant tannins could be useful as complements in mosquito species control programs.

According to Appel (1994), these changes may be associated with the process of activation of phenolic compounds through oxidation, which form free hydroxyl radicals that have a toxic action and are responsible for disrupting membrane integrity and causing metabolic disorders in the intestinal epithelium of insects. However, the changes that occur in the columnar cells are not only caused by allelochemicals and can also be triggered by other agents, as reported by Kakamand et al. (2008) who studied the effects caused by three chemical insecticides on the midgut *of Aphis mellifera*, and observed that the columnar cells showed changes similar to those described in this work. Levy (2005) noted that the columnar cells of the midgut *of A. gemmatalis* susceptible and resistant to *Ag* MNPV showed alterations in their structure after virus infection.

In addition to the intense vacuolization and cytoplasmic voids observed in the columnar cells, some treatments such as Dowling and IAC 100 in susceptible larvae showed total discarding of columnar cells into the intestinal lumen. This may be associated mainly with the process of cell renewal, in which altered or damaged cells are discarded into the intestinal lumen (BROOKS et al., 2002; KAKAMAND et al., 2008), 2008) and which here may be correlated with the insect's defense mechanism against its aggressor, in this case the high concentration of genistein present in the Dowling genotype (4.29 x 10^{-2} mg\g) and in the treatment with the IAC 100 genotype, the synergistic effect of the flavonoids rutin and genistein (R= 9.72x10^{-2} mg/g and G= 1.42x10^{-2} mg/g).

The formation of cytoplasmic protrusions being released from the epithelium was evident in all treatments in both larval populations and can be considered a normal physiological process of epithelial renewal. What caught our attention was the increase in these protrusions as the concentration of genistein and rutin present in the genotypes varied. In the BRS 257 treatment, there were few cytoplasmic protrusions formed or released towards the intestinal lumen, both in susceptible and resistant larvae to *AgMNPV*, while in the treatments with the other genotypes, even those with the lowest dosage of rutin and\or genistein, or even with no rutin at all, as in the case of the BR16 and Dowling genotypes, a large number of protrusions were observed, with the same response of alteration for both types of larvae.

The cytoplasmic protrusions do not yet have a well-defined function, but according to Anderson and Harvey (1966) these protrusions may be associated with columnar cells and are therefore related to the process of apocrine secretion. According to De Priester (1971), these protrusions may be involved in the process of cell degeneration that occurs during moulting or ecdysis and in the metamorphosis of insects, as observed in *Calliphora erytrocephala*. In addition to these characteristics, the process of protrusion formation can be related to the action of microorganisms or the ingestion of harmful products, where the epithelial cells seek out this formation process in order to "eliminate or expel" these agents from inside the cells (BROOKS et al., 2002). According to Martoja and Ballan-Dufrançais (1984), this process is characterized ultrastructurally by the vesiculation of the rough endoplasmic reticulum, the disappearance of the Golgi complex, an increase in the number and volume of autophagic vacuoles and also the thinning of the cytoplasm.

This process of forming and expelling protrusions has been observed by several authors in various insect species, such as *Helicoverpa armigera* (BROOKS et al., 2002); *Aedes aegypti* (ARRUDA et al., 2003); *Apis mellifera* (KAKAMAND et al., 2006); *Plodia interpunctella* (RHARRABE et al., 2007); *Spodoptera frugiperda* (CORREIA et al., 2009); *Alabama argilacea* (SOUSA et al., 2009).

The specific analysis of the alterations presented by the different genotypes shows that the columnar cells in both larval populations, both susceptible and resistant to *Ag* MNPV, when their cytoplasmic regions are considered, the greatest alterations are observed, the apical and middle regions, while the basal region only showed strong changes when treated with the Dowling genotype, which contains only the isoflavonoid genistein, but at a high concentration, i.e. 27 times the initial concentration of genistein.

Few studies have described the effects of genistein concentration on cells or tissues, such as those by Dang, 2009, who looked at the effects on adipocytes and the influence on obesity, and DI et al, (2008) who worked with light microscopy to determine the effects of genistein on cell morphology and also made use of calorimetric assay and cell proliferation antigen (PCNA immunocytochemistry); flow cytometry (to quantify cells in S phase and in apoptosis); fluorometric assay and confocal microscopy to detect caspase-3 and apoptotic activity, respectively. They observed that genistein at low concentrations (1 mg/mL) stimulated proliferation in one cell type (UtLM cells), increased PCNA labeling and the percentage of cells in S-phase and at higher concentrations (10 mg/mL) impaired morphology, significantly inhibited proliferation, decreased PCNA labeling, increased caspase-3 activity and induced apoptosis in all cell types and concluded that the effects of genistein are dependent on its concentration.

Another focal point that was altered was the striated edge of the columnar cells, which showed a strong degree of degeneration in the Dowling, PI 229358 and IAC 100 treatments in the *AgMNPV-resistant* larvae, while in the susceptible larvae only the Dowling and PI 229358 genotypes showed the same alteration.

The striated edge of the columnar cells has microvilli with the presence of a glycocalyx and as the columnar cells show alterations in the apical and middle regions, these are the regions that preferentially contain organelles such as rough endoplasmic reticulum, lysosomes, pinocytosis vesicles, elements of the cytoskeleton, Golgi complex, mitochondria and secretory granules (LEVY, 2005). All this is also reflected in the structure of the microvilli, altering them and also altering the glycoprotein coating of the glycocalyx, causing it to lose its own capacity for adhesion and recognition. This loss of function of the microvilli and also of the glycocalyx can interfere with the adhesion of AgMNPV viruses, since the virions cross the peritrophic membrane, considered to be the first protective barrier against the action of pathogens, through its pores, fuse to the membranes of the epithelial cells of the midgut, after recognizing specific receptors, starting the replication cycle (FLIPSEN et al. 1995, BARRET et al. 1998), in the columnar cells and transmitting the infection to other tissues (FLIPSEN et al., 1995; VOLKMAN, 1997; CASTRO et al., 1999; MOSCARDI; SOUZA, 2002). This is due to the disorganization of the microvilli, the probable alteration in the production of glycoproteins in the glycocalyx, and even alterations in the apical and middle regions of the columnar cells, with structural and functional impairment of the organelles, as reported by Martoja and Ballan-Dufrançais (1984) lead us to believe that they may have broken the caterpillar's resistance to its biological control, the *Ag* MNPV virus, allowing the virus to penetrate its cells and continue the viral replication process.

Although the goblet cells in the midgut of susceptible and resistant larvae did not show any major morphological changes in relation to the ingestion of the different genotypes, we did observe some changes in relation to the turgidity of the chamber and also in relation to the number of cells, These are apparently hypertrophic, with a slight increase in cell volume and perhaps an increase in the number of cells. However, this data needs morphometric and histometric analysis to corroborate our visual observations. These changes were observed in both susceptible and resistant larvae fed the Dowling and PI 227687 genotypes, which had the highest concentrations of genistein and rutin, respectively.

As for the regenerative cells, we described the changes observed in the midgut of susceptible larvae fed the Dowling and PI 227687 genotypes and in resistant larvae fed the IAC 100 genotype as strong. In fact, what caught our attention was the fact that these cells were apparently increased in number compared to the controls of each larval type. However, morphometric studies are needed for a more critical evaluation. The regenerative cells of the Dowling genotypes in susceptible larvae and the IAC 100 genotypes in resistant larvae showed large groups of cells, the so-called "regenerative cell nests". These regenerative cell nests are related to the process of epithelial renewal, which consists of the replacement of eliminated cells by others developed from these nests (LEHANE; BILLINGSLEY, 1996; CHAPMAN, 1998; CAVALCANTE; CRUZ-LANDIM, 1999).

This fact can be observed in *Melipona quadrifasciata anthidioides*, in which these nests were located at the base of the epithelium of the midgut of queen bees and physogastric queens, (FRANÇA et al., 2006); and in *Atelura formica* where around 26 to 30 nests of regenerative cells were found, where some were found in cell differentiation (ROST- ROSZKOWSKA; VILIMOVA; CHAJEC; 2010).

The susceptible larvae treated with the BR 16 genotypes showed moderate changes and the resistant larvae treated with the BR 16 and PI 227687 genotypes showed a slight increase in the number of regenerative cells. In this case, we are once again struck by the differences in the concentration of genistein, which, at the lowest concentration tested, promoted alterations and perhaps induced multiplication in these cells aimed at their cell replacement function. This fact was observed in *Aedes aegypti* by Arruda et al (2003), who found an increase in these cells. Cruz-Landim et al (1996) related this to the length of the insect's biological cycle, where in longer cycles these cells are found in small quantities, but when the cycle is short, regenerative cells are found in larger quantities.

The peritrophic membrane only underwent morphological changes in both susceptible and resistant larvae in two treatments. The susceptible larvae suffered changes in the peritrophic membrane when fed the IAC 100 and PI 227687 genotypes (higher rutin concentration). In fact, it was not possible to observe the presence of the membrane, which we considered to be an indication of strong alteration. In the *AgMNPV-resistant* larvae, the membrane was moderately altered for the PI 227687 genotype, being quite fragmented, while for the PI 229358 genotype, as we didn't even see traces of the peritrophic membrane, we considered it to be the maximum degree of alteration, i.e. strong alteration. Our results are partially in agreement with those obtained by Salvador (2008), who reported that the peritrophic membrane was not altered in any treatment and justified that the integrity of the peritrophic membrane is probably related to the absence of lipids in its constitution, since the action of flavonoids, especially rutin, is associated with lipid metabolism (LIMA et al., 2003; PIUBELLI, 2004).

With regard to the treatments that showed alterations in the peritrophic membrane, this may be associated with the disorganization that the columnar cells suffered in their cytoplasm, which prevents the production and release of the secreted products, as well as the quality of the product to be used and in this case, we remember that the components secreted for the formation of the peritrophic membrane are produced by the goblet cells. In addition, the intense release of cytoplasmic protrusions with cellular content can prevent the membrane from being organized in the structural way it needs to be in order to perform its function.

The muscular layer that externally covers the digestive tube, inserting it and attaching it to the haemocele, also showed alterations in both the susceptible and resistant larvae treated with the PI 229358, IAC 100 and PI 227687 genotypes, and the damage was considered to be intense, since even at specific points, the layers were disorganized and fragmented, thus affecting the entire organizational balance of the digestive tract in these locations, as well as metabolic exchanges with the hemolymph and other organs, such as the fat body, generating a compromising fact for the functional integrity of the midgut of the insects tested, which could lead to them not continuing the cycle and even dying.

Thus, the results show that all genotypes caused changes in both larval populations, suggesting that the effects caused by the flavonoid (rutin) together with the isoflavonoid genistein promote major morphological changes in the midgut in both larval populations of *A. gemmatalis*, and that the isoflavonoid genistein also causes changes in the midgut on its own.

6 CONCLUSIONS

The midgut of *Anticarsia gemmatalis* larvae susceptible and resistant to *Ag* MNPV showed major morphological changes in the cells of the epithelial layer for all treatments. These changes were punctual and occurred mainly in the columnar cells, which showed intense cytoplasmic vacuolization, the presence of cytoplasmic voids and an increase in the protrusions released towards the lumen, sometimes containing cell nuclei. Changes were also observed in the peritrophic membrane and muscle layer for some genotypes, all of which contain rutin and genistein in their leaf chemical composition.

In general, all the treatments caused changes in the midgut, with greater or lesser intensity depending on the tissue. It should be noted that the genotypes that caused the greatest changes in susceptible larvae was PI 227687, and in resistant larvae the genotypes responsible for the greatest changes were IAC 100 and PI 227687.

Thus, these genotypes, especially 227687, can be recommended for planting, as they have resistance characteristics to *A. gemmatalis* larvae, as well as being able to break down the resistance of soybean caterpillars to their biological control, the multiple nucleopolyhedrovirus *Anticarsia gemmatalis* (*Ag* MNPV).

REFERENCES

ABOT, A.R.; et al. Development of resistance by *Anticarsia gemmatalis* from Brazil and the United States to a nuclear polyhedrosis virus under laboratory selection pressure. **Biological Control**, v.7, n. 1, p.126-130, aug. 1996.

ANDERSON, E; HARVEY, W. Active transport by cercopia midgut. **The journal of Cell Biology**. V.31,n.1, p.107-134, Oct. 1966

APPEL, H.M. Phenolics in ecological interactions: the importance of oxidation.

Journal of Chemical Ecology, v.19, n.7, p.1521-1551, jul.1993.

ARRUDA, W.; OLIVEIRA,G.M.; SILVA, I.G. Ultrastructural study of the toxicity effect of *Magonia pubescens* (ST.HILL) extract on the mesenterium of *Aedes aegypti* (L.) larvae (Diptera: Culicidae). **Revista de Patologia Tropical**, v.37, n.3, p.255-267, jul.-set. 2008.

BARRET,J.W. et al. Concomitant primary infection of the midgut epithelial cells and the hemocytes of *Trichoplusia ni* by *Autographa californica* Nucleopolyedrovirus.

Tissue; Cell, [S.l.] v.30, p.602-616, 1998.

BILLINGSLEY, P.F.; LEHANE, M.J. Structure and ultrastructure of the insect midgut.

In: LEHANE, M.J., BILLINGSLEY, P.F. (Eds). **Biology of the Insect Midgut**.

London: Chapman; Hall, 1996, p.3-30.

BROOKS, M.E. et al. Infection of its lepidopteran host by the *Helicoverpa armigera stunt virus* (Tetraviridae). **Journal of Invertebrate Pathology**, v. 80, p.97-111, 2002.

CERRI, P., SASSO-CERRI, E. Staining methods applied to glycol methacrylate embedded tissue sections. **Micron**, v.34, n.8, p.365, 2003.

CASTRO, M.E.B. et al. Molecular biology of baculoviruses and their use in pest control in Brazil. **Pesquisa Agropecuària Brasileira**, Brasilia, v.34, n. 10, p. 17331761, Oct. 1999.

CAVALCANTE, V.M.; CRUZ-LANDIM, C. Types of cells present in the midgut of the insects: a review. **Naturalia**, Rio Claro, v.24, p. 19-40, 1999.

CHAPMAN, R.F. **The Insects: structure and function**. 4.ed. Cambridge: Harvard University Press, 1998.

COMPANHIA NACIONAL DE ABASTECIMENTO. **Agricultural Indicators**. Available at: http://www.conab.gov.br/conabweb/IA-agos09.pdf . Accessed on October 6, 2009.

CORREIA, A.A.et al. Morphology of the alimentary canal of caterpillars *of Spodoptera frugiperda* (J.E. Smith) (Lepidoptera: Noctuidae) fed on leaves treated with Nim. **Neotropical Entomology**, Vacaria, v.38, n.1, p.83-91, jan-feb. 2009.

CRUZ-LANDIM, C.; SILVA de MORAES, R.L.M.; SERRÃO, J.E. Ultrastructural aspects of epithelial renewal in the midgut of adult worker bess (Hymenoptera:Apidae). **Journal of Computational Biology**, v.1, p. 29-40, 1996.

DANG, Z.C Dose-dependent effects of soy phyto-oestrogen genistein on adipocytes: mechanisms of action. **Obesity Reviews**, v.10, n. 3, 342-349, 2009.

DEGRANDE, P.E.; VIVAN, L.M. Pragas da Soja. In: **Soybean Research Bulletin**. Rondonópolis: Fundaçâo MT, 2006. p.153-169.

DE PRIESTER, W. Ultrastructure of the midgut epithelial cells in the fly *Callifora erythrocephala*. **Journal of Ultrastructure Research.** New York v.36, n.5, p. 783805, set. 1971.

DI, et al. A low concentration of genistein induces estrogen receptor-alpha and insulin-like growth factor-I receptor interactions and proliferation in uterine leiomyoma cells. **Human Reproduction**, [s.l], v.23, n.8, p.1873-1883, maio, 2008.

EISEMANN, C.H.; BINNINGTON, K.C. The peritrophic membrane : its formation, structure, chemical composition and permeability in relation to vaccination against ectoparasitic arthropods. **International Journal for Parasitology**, London, v.24, n.1,p.15-26, feb. 1994.

EMBRAPA. **Production Systems: Technology and soybean production in central Brazil 2006.** 1.ed. Londrina: Embrapa-CNPSo, 2005.

FELTON, G.W. et al. Activation of plant foliar oxidases by insect feeding reduces nutritive quality of foliage for noctuid herbivores. **Journal of Chemical Ecology**, California, v.15, n.12, p.2667-2694, 1989.

.; DUFFEY, S.S. Enzimatic antinutritive defenses of the tomato plant against insects.In: HEDIN, P. (Ed.) **Naturally occurring pest bioregulators.** Washington: American Chemical Society, p.166-197, 1991.

FLIPSEN, J.T.M.; et al. Passage of *Autographa californica* nuclear polyhedrosis virus through the midgut epithelium of *Spodoptera exigua* larvae. **Virology**, New York v.208, p.328-335, 1995.

FRANÇA, A.A.P. et al. Regenerative cells in the midgut of *Melipona quadrifasciata anthidiuides* (Hymenoptera, Apidae Melliponini): A comparative study of workers and quens. **Brazilian Journal Morphological Science**, v.23, n. 3-4, p.401-404, 2006.

FUGI, C.G.Q.; et al. Biology of *Anticarsia gemmatalis* on soybean genotypes with different degress of resistance to insects. **Scientia Agricola**, Piracicaba, v.62, n.1, p.31-35, jan./feb. 2005.

GALLO, D.; et al. **Entomologia agricola**. Piracicaba: FEALQ, 2002. v.10.

GAZZONI, D.L. HÜLSMEYER,A.; HOFFMANN-CAMPO, C.B. Effect of different doses of rutin and quercetin on the biology of *Anticarsia gemmatalis* Hübner, 1818 (Lep., Noctuidae). **Revista Agropecuaria Brasileira**, Brasilia, v.32, n.7, p. 673-681, jun./jul. 1997.

HOFFMANN-CAMPO, C.B. et al. **Mass rearing of the soybean caterpillar (*Anticarsia gemmatalis*).** Londrina: EMBRAPA/CNPSo, 1985, 23p.

HOFFMANN-CAMPO,C.B.; et al. Mechanisms of resistance of soybean genotypes: non-preference test for *Anticarsia gemmatalis* Hübner, 1818 (Lep.:Noctuidae). **Pesquisa Agropecuària Brasileira**, Brasilia, v.29, n.4, p. 513-519, Apr. 1994.

HOFFMANN-CAMPO, C.B.; **Role of the flavonoids in the natural resistance of soyabean to *Heliothis virescens* (F.) and *Trichoplusia ni* (Hübner)**. 1995.

Dissertation (PhD in Botany) - 165 p. The University of Reading, Reading.

HOFFMANN-CAMPO, C.B.; et al. **Soybean pests in Brazil and their integrated management.** Londrina: EMBRAPA-CNPSo, 2000.

HOFFMANN-CAMPO, C.B.; et al. Pre-ingestive and and post-ingestive effects of soybean extracts and rutin on *Trichoplusia ni* growth. **Entomologia Experimentalis et Applicata**, Amsterdam, v.98, n.2, p.181-194, feb. 2001.

KAKAMAND, F.A.K.H.; MAHMOUD,T.T.; AMIN,A.B.B. The role of three insecticides in disturbance the midgut tissue in honey bee *Apis mellifera* L. workers. **Journal Dohuk University**, Kurdistan v.11, n.1, p. 144-151, May. 2008.

KUBO, I.; HANKE, F.G. Chemical methods for isolating and identifying phytochemicals biologically active in insects. In: MILLER, J.R.; MILLER, T.A. (Eds) **Insects plants interactions**. New York: Spring Verlog. 1986 p.225-249.

LARA, F.M. **Principios de resistência de plantas a insetos**. Sao Paulo: icone, 1991.

LEVY, S.M. et al. The larval midgut of *Anticarsia gemmatalis* (Hübner) (Lepidoptera: Noctuidae): light and electron microscopy studies of the epithelial cells. **Brazilian Journal of Biolpogy**, Sao Paulo, v.64, n. 3b, p.1-6, aug. 2004.

LEVY,S.M. et al. Morphology of the midgut of *Anticarsia gemmatalis,* susceptible and resistant to nuclear polyhedrosis virus (*AgMNPV*), in control and virus-infected larvae. In: SIMPÒSIO DE CONTROLE BIOLÒGICO, 4., 2005, Recife.

Abstracts... Recife: FIOCRUZ, 2005. p.188.

LIMA, L.R.P. et al. Acute toxicity of rutin and bixin from *Bixa orellana*. **Acta Farm Bornaerense**, v.22, n.1, p.21-26, 2003.

LIMA, L.R.P. et al. Determination of lipase activity in the presence of morin, naringenin, naringin and rutin. **Revista de Ciência e Agrotecnologia**, Lavras, v. 23, n. 3, p. 626-631, jul./set. 1999.

LOPES, R.M. et al. Effect of different doses of rutin on the concentrations of albumin, creatine, urea and uric acid in the serum of rabbits. **Revista Brasileira de Farmacia**, Rio de Janeiro, v. 86, n.1, p. 35-38, jan. 2005.

MARTOJA, R.; BALLAN DUFRANÇAIS, C. The ultrastructure of the digestive and excretory organs. In: KING, R.C.; AKAI, H. (Eds). **Insect Ultrastructure.** New York: Plenum Press,1984. v.2, p.119-261.

MOSCARDI, F.; CARVALHO, R.C.Z. Consumption and utilization of soybeans by *Anticarsia gemmatalis* Hüb. (Lepidoptera: Noctuidae) infected at different larval stages by its nuclear polyhedrosis virus. **Anais Sociedade Entomològica do Brasil**, Londrina, v.22, n.2, p.267-280, jun.1993.

MOSCARDI, F. Assessment of the application of baculoviruses for control of lepidoptera. **Annual Review Entomology**, v.44, p.257-289, 1999.

MOSCARDI, F.; SOUZA, M. L. de. Baculovirus for pest control. **Biotecnologia Ciência e Desenvolvimento**. n.24, p. 22-29, jan./feb. 2002.

OMYILAGHA, J.C.; GROTEWOLD, E. The biology and structural distribution of surface flavonoids. **Recent Research developments in plant science**, Columbus Ohio, v.2, p. 53-71, 2004.

PAUTOU MP. et al. Toxicity of vegetable tannins on crustaceae associated with Alpine mosquito breeding sites. **Ecotoxicology and Enviromental Safety**, v. 47, p.323-332, 2000.

PIUBELLI, G.C. **Bioactivity of soybean genotypes resistant to *A. gemmatalis* Hübner (Lepidoptera: Noctuidae) and interactions of their chemical substances with natural enemies**. 2004.Thesis (Doctorate in Biological Sciences - Entomology), Universidade Federal do Paranâ, Curitiba, Paranâ. 152p.

PIUBELLI, G.C.et al. Are chemical compounds important for soybean resistance to

Anticarsia gemmatalis ? **Journal of Chemical Ecology**, California, v.31, n.7,p.1509- 1525, jul. 2005.

REY D., PAUTOU, MP; MEYRAN, J.C. Histopathological effects of tannic acid on the midgut epithelium of some aquatic diptera larvae. **Journal of Invertebrate Pathology**, v.73, p.173-181, 1999.

RHARRABAE, K et al. Bioinsecticidal effect of hamaline on *Plodia interpunctella* development (Lepidoptera: Pyralidae). **Pesticide Biochemistry and Physiology**, v.89, p. 137-145, May 2007.

ROST-ROSZKOWSKA, M.M.; VILIMOVA, J.; CHAJEC, L. Fine structure of the midgut epithelium *of Atelura formicaria* (Hexapoda: Zygantoma: Ateluridae), with special reference to its regeneration and degeneration. **Zoological Studies**, v. 49, n.1, p. 10-18, 2010.

SALVADOR, M.C. **Effect of soybean genotypes and flavonoids on the biology and midgut of *Anticarsia gemmatalis***. 2008. Dissertation (Master's Degree in Agricultural Entomology). Faculty of Agricultural and Veterinary Sciences - UNESP Jaboticabal Campus, Jaboticabal.

SILVA, H.H.G.S. et al. Larvicidal activity of tannins isolated from *Magonia pubescens* St. Hil. (Sapindaceae) on *Aedes aegypti* (Diptera: Culicidae). **Revista da Sociedade Brasileira de Medicina Tropical**, Uberaba, v.37, n.5, p. 396-399, set./out. 2004.

SILVA, R.R. et al. Hypolipidemic effect of the flavonoids naringin and rutin. **Archivos Latinoamericanos de Nutrición**, Caracas, v.51, n.3, p. 258-264, Apr. 2001.

SILVA, R.R. et al. Effect of flavonoids on arachidonic acid metabolism. **Medicina**, Ribeirao preto, v.35, p. 127-133, apr./jun. 2002.

SOSA-GÓMEZ, D.R. et al. **Manual of identification of insects and other invertebrates of the soybean crop.** Londrina: EMBRAPA-CNPSo, 2006.

SOUSA, M.E. et al. Ultrasctructure of the *Alabama argilacea* (Hübner) (Lepidoptera:Noctuidae) midgut. **Micron**, v.40, p.743-749, 2009.

STAMP, N.E.; SKROBOLA, K.L. Failure to avoid rutin diets results in altered food utilizatin and reduced growth rate of *Manduca sexta* larvae. **Entomologia Experimentalis et Applicata**, Netherlands, v. 68, n.2, p.127-142, aug. 1993.

TERRA, W.R.; FERREIRA, C. The digestion of insects. **Ciência Hoje,** v.12, p.28-38, 1991.

TERRA, W. R. The origin and functions of the insect peritrophic membrane and peritrophic gel. **Archives of Insect Biochemistry and Physiology**, New York, v.47, n. 2, p.47-61, jun. 2001.

TOLEDO, T., et al., Effects of differents rutin dosages on chloride, calcium and phosphorum levels serum of rabbits. **Revista Chilena de Nutrición**, v.30,n.3, p.263270, dec. 2003.

VALLE, G.E.; LOURENÇÂO. A.L. Resistance of Soybean Genotypes to *Bemisia tabaci* (Genn.) Biotype B (Hemiptera: Aleyrodidae). **Neotropical Entomology**, Vacaria, v.31, n.1, p.285-295, apr./jun. 2002.

VOLKMAN, L.E. Nucleopolyhedrovirus interactions with their insect hosts. **Advances in Virus Research**, v.48, p.313-348, 1997.

WANG, P.; GRANADOS, R.R. Molecular structure of the peritrophic membrane (PM): identification of potential PM target sites for insect control. **Archives of Insect Biochemistry and Physiology**, New York, v.47, n.2, p.110-118, jun.2001

4 ARTICLE B - EFFECTS OF FEEDING WITH INSECT-RESISTANT SOYBEAN GENOTYPES ON THE EPITHELIAL SURFACE OF THE MIDDLE GUT OF SOYBEAN LAGARTS SUSCEPTIBLE AND RESISTANT TO THE NUCLEAR POLYEDROSIS VIRUS *AGMNPV*. ELECTRONIC SCREEN MICROSCOPY STUDY

SUMMARY

The soybean caterpillar *Anticarsia gemmatalis* is one of the main pests of soybeans and is responsible for losses in final grain production. It is controlled by applying biological insecticides such as the nucleopolyhedrovirus *Ag* MNPV and resistant cultivars or plants that avoid damaging the environment. The use of resistant plants causes negative physiological changes in insects due to the presence of chemical substances such as allelochemicals like flavonoids (rutin and genistein). The action of pathogens such as AgMNPV and flavonoids takes place in the midgut, which is the main access route for these substances. The aim of this study was to verify the morphological changes in the surface of the epithelial cells of the MI caused by the ingestion of rutin (R) and genistein (G), at their maximum concentrations, in *A. gemmatalis* larvae susceptible and resistant to *AgMNPV*. These larvae came from Embrapa Soja/LondrinaPR and were treated with "in natura" soybean genotypes containing rutin and genistein: BRS 257 (control); Dowling (G= 0.0429) and PI 227687 (R=0.3682 and G= 0.0122mg⁄g). The MIs of the 4th instar larvae were fixed, processed and analyzed using a scanning electron microscope. The susceptible and resistant larvae treated with the control genotype had columnar cell surfaces with no morphological changes, but in the treatments with the Dowling and PI227687 genotypes, changes were evident in both the microvilli and the cytoplasmic protrusions in both larval populations. There was a high concentration of cytoplasmic protrusions of varying shapes and sizes released towards the intestinal lumen. These protrusions were more intense in the PI 227687 treatment, both in susceptible and resistant larvae compared to the Dowling treatment. With regard to the quantity of protrusions released between the two types of larvae, the resistant larvae showed a greater quantity when compared to the susceptible larvae.

keywords: Soybean caterpillar. Columnar cells. Protrusions. Flavonoids. Resistance. Rutin. Genistein.

1 INTRODUCTION

The soybean caterpillar, *Anticarsia gemmatalis* (Hübner) (Lepidoptera: Noctuidae) is considered one of the main defoliating pests of the soybean crop in Brazil, and its behavior and biology have been extensively studied (MOSCARDI; CARVALHO, 1993; HOFFMAN et al., 2006, MENDONÇA et al., 2009).

In order to reduce production costs and unfavorable environmental impacts, great efforts have been made in Brazil to develop methods of controlling *A. gemmatalis* and other soybean insect pests. The bioinsecticide based on the virus *AgMNPV* (*Anticarsia gemmatalis* multiple nucleopolyhedrovirus), specific for this pest, has been used as a biological control since the 1980s (MOSCARDI; CARVALHO, 1993; MOSCARDI, 1999; ABOT et al., 1996). Infection with *AgMNPV* begins when the caterpillars ingest the viral polyhedra together from the soybean leaves; the virions are released into the alkaline lumen of the midgut, pass through the peritrophic membrane and fuse with the membrane of the epithelial cells, after recognizing specific receptors, starting the replication cycle (FLIPSEN et al., 1995). Another highly recommended alternative technique is the use of plant resistance, which has, among other advantages, practicality of use and safety (HOFFMANN-CAMPO et al., 2001).

Plant resistance to insects is mainly due to chemical agents, i.e. substances (allelochemicals) present in host plants, such as alkaloids, flavonoids, terpenoids and steroids (KUBO and HANKE, 1986). Although the chemical importance of these substances has been recognized since the early 1950s, only recently has their use been considered and they are still little studied (KOGAN, 1986).

In the food preference test with *A. gemmatalis*, the genotypes BR82-12547, IAC74-2832, PI 227687, PI 229358, PI 274454 were rejected by the caterpillars (HOFFMANN-CAMPO et al., 1994). The effect of various soybean genotypes on the biology of *A. gemmatalis*, as well as food consumption and utilization was estimated by Machado et al., (1999); in which caterpillars fed on the resistant cultivar IAC-100 showed longer larval development and lower weight compared to caterpillars fed on other genotypes tested.

Phenolic compounds, such as rutin and chlorogenic acid, are considered models in studies of plant anti-herbivory defense. The effects of flavonoids can vary depending on the species of insect, and frequently, monophagous and oligophagous insects such as *A. gemmatalis* can use them to recognize their host plants (HARBORNE, 1991). Although the aglycone quercitin and rutin, one of its glycosides, have caused increased mortality and lengthened the larval period of *A. gemmatalis* (GAZZONI et al., 1997), the nutritional and post-ingestive effects of flavonoids still require further investigation (HOFFMANN-CAMPO, 1995; PIUBELLI et al., 2005).

A complete understanding of the effects of flavonoids such as rutin and genistein on a specialized pest that is the main defoliator of soybeans, such as *A. gemmatalis*, as well as their behavior and nutrition, is of great importance for the success of chemical breeding and pest control programs (PIUBELLI et al., 2005).

Thus, pest control programs aimed at reducing the resistance of *A. gemmatalis* larvae to AgMNPV have used chemical substances found in extracts of soybean leaves, which promote the mortality of larvae resistant to the virus compared to susceptible larvae (HOFFMANN-CAMPO, 1995; PIUBELLI et al., 2005).

As the midgut of insects is the main route of access and action of both bioinsecticides and flavonoids, there is a need to study the changes caused by these substances.

The digestive tract of insects is an efficient physical and chemical barrier against the action of microorganisms that may be ingested along with the food. In general, the alimentary canal comprises three regions: the anterior, middle and posterior intestines, with the middle intestine being the main region, essential for the digestion and absorption of food, as well as chemical and biological phytosanitary products (TERRA; FERREIRA, 1991; CHAPMAN, 1998).

Morphologically, the midgut *of A. gemmatalis* larvae is made up of several morphofunctionally distinct structures. From the lumen to the periphery of the organ, lining the epithelium, is a thin acellular layer called the peritrophic membrane which divides the lumen into two spaces, the ectoperitrophic and the endoperitrophic. The epithelium is of the pseudostratified columnar type with at least four distinct cell types - columnar, goblet, regenerative and endocrine cells. Below the epithelium is the muscular layer made up of transverse and longitudinal muscle bundles (LEVY et al., 2007; 2008).

Some studies on lepidopteran larvae suggest that the different cell types found in the epithelium of the midgut show morphological and functional variations along its length, such as those reported by Pinheiro et al. (2003), Pinheiro et al. (2008), Levy et al. (2009), Cruz et al. (2010, in press) who worked with Lepidoptera larvae.

The main objective of this work was to study the changes caused to the surface of the epithelial cells of the midgut *of A. gemmatalis* larvae susceptible and resistant to *AgMNPV*, fed on leaves of different soybean genotypes containing different concentrations of these flavonoids.

2 MATERIAL AND METHODS

The BRS 257 soybean genotype was used as a control, as well as the Dowling and PI 227687 insect resistance genotypes. The concentration of the flavonoids rutin and genistein present in these genotypes was (0.0 and 0.0); (0.0 and 0.0429); (0.3682 and 0.0122), respectively for rutin and genistein. These genotypes were sown and cultivated in a greenhouse at Embrapa Soja Londrina-PR under ideal conditions of temperature, pressure, humidity and planting.

The leaves were collected, taken to the laboratory and washed in a 5% sodium hypochlorite solution; rinsed in distilled water; dried on filter paper and placed in gerboxes with *A. gemmatalis* eggs, susceptible and resistant to *AgMNPV*. These eggs were supplied by the Entomology Laboratory of EMBRAPA Soja, Londrina/PR, and came from susceptible and resistant insects (Lethal Concentration - CL_{80} in each generation), reared on an artificial diet (HOFFMANN-CAMPO et al., 1985) under controlled temperature (2527°C), photoperiod (14h light/10h dark) and with 80% relative humidity. After the larvae hatched, they were repotted at the 2nd instar stage and kept at 25 larvae per germbox, to facilitate their maintenance. The soya leaves were changed every 12 hours to prevent food shortages.

A. gemmatalis larvae susceptible and resistant to AgMNPV were immobilized and anaesthetized when they reached the fourth instar (10-12 days old) by cooling them in the freezer for about 5 minutes and dissected under a stereoscope with saline solution for insects (1.80 g NaCl; 1.88 g KCl; 0.16 CaCl; 0.004 g $NaHCO_3$ and distilled water-q.s.p 100 mL). The larvae were pinned at the ends by entomological pins in a Petri dish filled with paraplast and, with the help of ophthalmic scissors, a logitudinal incision was made in the ventral region, from the first thoracic segment to the penultimate abdominal segment, with the cuticle folded back laterally. The IMs were identified and sectioned in half, transversely and lengthways, and laterally folded to expose the internal surface. For analysis under the scanning electron microscope, 10 digestive tubes were collected per treatment for both larval populations.

The collected midguts were fixed in a solution of

2.5% glutaraldehyde in 0.1M sodium phosphate buffer pH 7.3 for 24 hours and washed in the same buffer; powders fixed in 1% osmium tetroxide in 0.1M sodium phosphate buffer pH 7.3 for 30 minutes; dehydrated in increasing series of ethyl alcohol (50 to 100%); critical point dried with liquid CO_2 and coated with 10 nm of gold.

Ultrastructural analysis of the material was carried out using a Quanta 200 (FEI) scanning electron microscope at the UEL Laboratory of Electronic Microscopy and Microanalysis (LMEM/UEL).

3 RESULTS

In susceptible larvae (LS) treated with the control genotype BRS 257, the midgut (MI) was made up of the anterior (AR), median (MR) and posterior (PR) regions, which were easily identifiable The surface of the AR of the MI showed variation in the morphology of the columnar cells (Fig. 4.1 A). In part, the surface of the cells was flattened, juxtaposed, of varying sizes, covered in discrete microvilli and lacking cytoplasmic protrusions (Fig. 4.1 A). In other portions, the cells were larger and bulging, of varying sizes, with many microvilli and some cytoplasmic protrusions (Fig. 4.1 B.). These protrusions were rounded, of varying sizes and had a rough irregular surface. We also observed the presence of holes of varying sizes in the apical portion of the cell and the release of secretion droplets.

The median region showed little variation in the surface of the columnar cells when compared to RA, with a slight increase in the concentration of rough cytoplasmic protrusions (Fig. 4.1B).

The posterior region showed cell surface morphology similar to the other two (RA and RM), but what was most striking was the intense secretion of apical cytoplasmic protrusions towards the ectoperitrophic space (Figs. 4.1C and D).

In resistant larvae (LR) treated with the same genotype, cytoplasmic protrusions were detected along the entire epithelial surface of the MI. The RA, RM and RP regions did not have clearly established boundaries, but showed columnar cells of varying diameters and heights with extensive and abundant microvilli. There were apical cytoplasmic protrusions of various sizes and shapes, most of which had a rough irregular surface, but there was a slight presence of other types of protrusions, which were rounded and had a smooth surface (Fig. 4.1 E). Several of these protrusions were visualized still being released from the cell and large holes were observed between the microvilli (Fig. 4.1F).

In the treatment with the Dowling genotype, in the LS, the IM showed only two identifiable regions, the RA and the RP (Fig. 4.2A). The RP showed a number of cytoplasmic protrusions, either rounded and smooth or with granules on its surface (Figs. 4.2 C and E). RA showed fewer protrusions, as well as a small number of perforations on the surface of the columnar cells (Figs. 4.2 A and B).

In addition to the large number of protrusions being released from the columnar cells (Fig. 4.2 C), there were variations in the shape and quantity of the microvilli, which were sometimes very long and free (Fig. 4.2 D) and sometimes shorter and concrescent (Fig. 4.2 D).

In the LR, it was difficult to identify the MI regions. The columnar cells showed surfaces of varying sizes, covered with concrescent microvilli (Fig. 4.2 E) and some nests of free microvilli with many protrusions (Fig. 4.2E). The protrusions were irregularly shaped, smooth or rough (Fig. 4.2 H) and rounded and smooth (Fig. 4.2 F).

In the LS treated with PI 227687 genotype leaves, the IM showed both anterior and posterior regions. In the AR, most of the columnar cells had a flattened surface, with short microvilli (Fig. 4.3 A), but to a small extent a convex surface could be seen, with short, concrescent microvilli and no protrusions (Fig. 4.3 B). The posterior region was quite distinct, with a large number of protrusions of varying shapes and surfaces (Fig. 4.3 B). Irregular cytoplasmic protrusions were identified (Fig. 4.3 B), with a rough and/or granular surface (Fig. 4.3 B) and rod-shaped protrusions with a smooth surface. In this region, the microvilli were concrescent and short (4.3 A and B)

In the LR, it was possible to observe the anterior and posterior regions of the IM, quite individualized, with the AR showing the surface of the columnar cells without protrusions, with varying sizes, juxtaposed and with few microvilli (Fig. 4.3 C), while the RP showed a large number of cytoplasmic protrusions being released into the lumen (Fig. 4.3 D). The cytoplasmic protrusions were mostly smooth rounded protrusions with small granules on their surface. Several protrusions were still being released from the surface of the columnar cells (Fig. 4.3 D), as well as holes in the surface of these cells after the cytoplasmic protrusions had been released. In addition to protrusions, droplets of secretion and short, concrescent microvilli were observed.

It was also possible to observe a third region, probably the MRI, which had an area with a large number of protrusions and another area with a very small number of protrusions.

4 PHOTOGRAPHIC DOCUMENTATION

A - Anterior region of the IM of LS. Flattened epithelial cells with discrete microvilli (>) and without microvilli (*); juxtaposed (Cj) covered with microvilli (>). Bar = 20µm.

B - Median region of the IM of LS. Bulging cells (Ab) with microvilli (>) and rough cytoplasmic protrusions (Pr). Bar = 20 µm

C - Posterior region of the IM of LS. Cells with a large number of smooth (Pl) and rough (Pr) cytoplasmic protrusions. Bar = 20µm.

D - Anterior region of LR IM. Juxtaposed cells covered with microvilli (>) Bar = 20 µm

E and F - Posterior region of LR IM. Cells with cytoplasmic protrusions (P) and microvilli (>) on the surface of the columnar cells (Mi). Note the holes (probably from the exit of the protrusions) between the microvilli (▼). Bar = 20 µm.

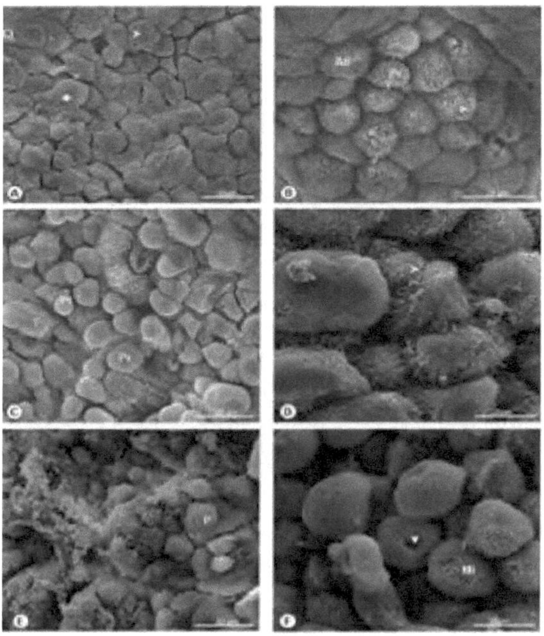

Figure 4.1 - Electron micrographs of the midgut of *AgMNPV-susceptible* and -resistant *A. gemmatalis* larvae treated with the BRS 257 genotype (control)

A - Anterior region (AR) of the IM of the LS with juxtaposed cells (Cj) and posterior region (RP) with protrusions (P). Bar = 50µm.

B - Anterior region of the IM of LS. Juxtaposed cells (Cj), with short, joined microvilli (Mic), cytoplasmic protrusions with granules (Pg) and secretory droplets (Se). Bar = 20 µm.

C and D - Posterior region of the IM of LS. Cells with numerous smooth-type cytoplasmic protrusions (Pl), short, joined microvilli (Mic) and cells with long, free microvilli (Mil). Bars C = 20µm and D = 10µm.

E and F - LR IM surface. (E) Anterior region with juxtaposed columnar cells (Cj) covered with short, concrescent microvilli (Mic). (F) Posterior region with columnar cells with short, conjoined microvilli (Mic) and smooth (Pl) and rough (Pr) cytoplasmic protrusions. Bar = 10µm.

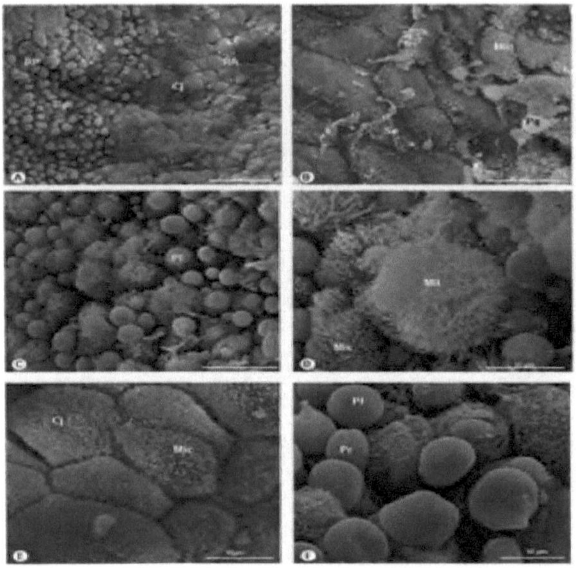

Figure 4.2 - Electron micrographs of the midgut of *AgMNPV-susceptible* and -resistant *A. gemmatalis* larvae treated with the Dowling genotype

A and B - Surface of LS IM cells. (A) Anterior region of LS IM. Juxtaposed columnar cells (Cj) covered with short and close (Mic) or long (>) microvilli. (B) - Posterior region of the IM of LS. Columnar cells covered with short, jointed microvilli (Mic) and granular cytoplasmic protrusions (Pg). Bar = (A) 20µm and (B) = 5 µm.

C and D - Surface of LR IM cells. (C) Anterior region with juxtaposed cells (Cj). (D) Posterior region of the LR IM. Columnar cells with a large number of smooth cytoplasmic protrusions (Pl). Bar = (C) 20µm and (D) = 50µm.

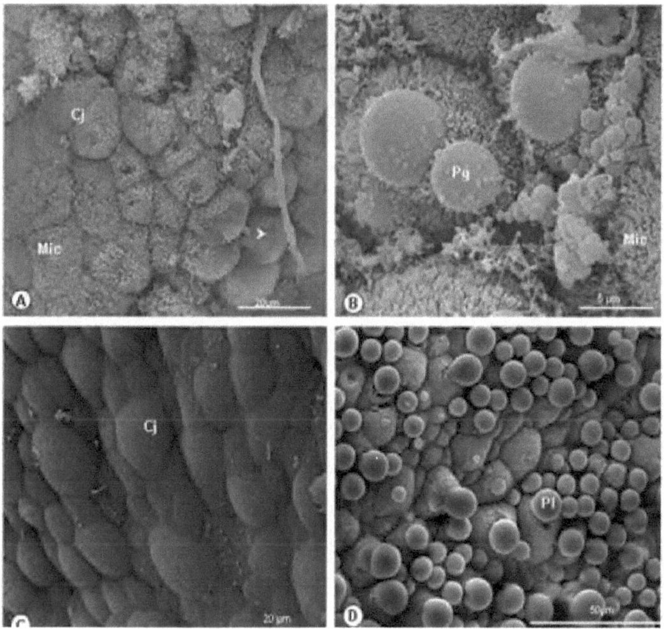

Figure 4.3 - Electron micrographs of the midgut of AgMNPV-susceptible and -resistant *A. gemmatalis* larvae treated with the PI 227687 genotype

5 DISCUSSION

The general morphology of the midgut wall in the population of *A. gemmatalis* larvae susceptible and resistant to AgMNPV treated with the control genotype BRS 257 is similar to that described for this same type of larva when treated with an artificial diet free of these flavonoids, as described by LEVY et al. (2008); CRUZ et al. (2010) and with the results described for the majority of Lepidoptera species (BILLIGSLEY; LEHANE, 1996, RIBEIRO et al., 1997; BROOKS et al., 2002; PINHEIRO et al., 2008).

In susceptible larvae, the differences in their surface made it possible to differentiate three distinct regions in the MI: anterior, middle and posterior. These regions were also identified in resistant larvae, but without clearly delimited boundaries. In general, the surface of the columnar cells varied in terms of size, height, shape, microvilli and cytoplasmic protrusions. There was a gradual increase in these protrusions from the anterior to the posterior region.

The cytoplasmic protrusions were the main structures that made it possible to differentiate the regions of the midgut. Apparently the intensity and shape of these protrusions increased according to the treatment to which the larvae were subjected.

The protrusions were rounded and smooth without or with granules (probably from the microvilli) and rough irregular ones, or even detached from the apical portion of the columnar cells, similar to those described by Pinheiro (2006); Pinheiro et al. (2008).

This variation in the shape of the protrusions was described by Jimenez and Gillian (1990) when they observed, using transmission electron microscopy, two types of protrusions that varied in the middle and posterior regions of the midgut *of Apis mellifera*. The authors reported that these granular structures were part of the merocrine secretion process, while those with a smooth appearance and cellular content were related to the holocrine secretion process. Levy (2005) described the variation in the shape of the protrusions, relating those with a rounded appearance as protrusions with cytoplasmic content and those with a rough appearance as protrusions whose content had already leaked into the lumen.

Many of these protrusions had small holes in the membrane on their surface, which were also observed by Pinheiro (2006) who reported these characteristics as interruptions in the membrane of the protrusions.

The BRS 257 genotype (control) promoted discrete protrusions in small quantities in both susceptible and resistant larvae, while susceptible and resistant larvae to *Ag* MNPV fed the Dowling and PI 227687 genotypes showed an increase in the concentration of these protrusions, which varied in size and shape along the midgut from the anterior to the posterior region. This fact was also observed in *Diatrea saccharalis* larvae by Pinheiro et al. (2008).

These protrusions have been described for many years, but their function has not been determined. Some authors have reported that these protrusions are associated with insect ecdysis, as described by De Priester (1971) for *Calliphora erythrocephala*, or even with the process of cellular degeneration resulting from epithelial renewal (ANDERSON, HARVEY, 1966; DE PRIESTER 1971; TERRA; FERREIRA, 1994) and the secretion of digestive enzymes (SANTOS, et al., 1984; WOOD; LEHANE, 1991; CRISTOFOLETTI et al., 2001).

However, in studies carried out on *Blatella germanica*, the secretion of enzymes was contested due to the reduced enzymatic activity found in the midgut, even though it had a high concentration of these proteins (JIMENEZ; GILLIAM, 1990).

The function of these protrusions can also be related to the defense process of the insects, which eliminate harmful substances or even damaged or infected cells into the intestinal lumen. In our studies, we observed that both susceptible and resistant *A. gemmatalis* larvae fed on Dowling and PI 227687 genotypes containing the highest concentrations of the flavonoids rutin and genistein showed intense production of protrusions in the midgut, unlike the control treatment.

The fact that protrusions are released as a defense system was observed by De Priester (1971) in *Calliphora erythrocephala*; Washbum et al. (2001) in *Helicoverpa zea* and *Heliothis virescens*; Brooks et al. (2002) in *Helicoverva armigera*; Arruda, Oliveira; Silva (2003) in *Aedes aegypti*; Pinheiro et al.

(2006) in *Diatrea saccharalis*; Khakamand, Mahmound, Amin (2008) *Apis mellifera.* These authors observed intense secretion after ingestion of harmful substances or infection of the individual, with elimination of the epithelium of diseased and damaged cells.

Thus, the fact that *A. gemmatalis* larvae show cytoplasmic protrusions varying in intensity between the susceptible and resistant populations is corroborated by data in the literature that relate this fact to the resistance process and also by Levy (2005), Cruz et al. (2010) who, when analyzing the midgut of *A. gemmatalis* larvae susceptible and resistant to *AgMNPV* using scanning electron microscopy and treated with an artificial diet, noticed that the concentration of protrusions in resistant larvae was higher.

The microvilli covering the surface of the columnar cells were found to vary in density and size, as described by Pinheiro et al (2008) in *Diatrea saccharalis*.

In the anterior region, the microvilli were found with little density and concrescence and in the posterior region they were longer and freer. One of the functions of the microvilli is to protect the columnar cell layer and the *Ag* MNPV virus has to cross these microvilli before penetrating the columnar cell membrane in order to start the virus replication process inside these cells (FLIPSEN et al., 1995; VOLKMAN, 1997).

Another variation observed among the microvilli was the presence of holes between them, which can be attributed precisely to the release of cytoplasmic protrusions that carried pieces of these structures on their surface.

Drops of secretion were also noted near the microvilli, especially in the Dowling treatment with resistant larvae. These drops seem to be associated with the release of digestive enzymes from the columnar cells into the lumen and are related to the microapocrine secretion process that occurs from the microvilli (SANTOS et al., 1984; WOOD, LEHANE 1991, CRISTOFOLETTI et al., 2001).

The absence of these microvilli on the surface of the columnar cells can facilitate the adhesion and penetration of foreign bodies and pathogens that can infect and cause disruption of the cells, thus accelerating the process of cell infection.

6 CONCLUSIONS

The midgut of susceptible and resistant *Anticarsia gemmatalis* larvae showed three distinct regions: anterior, middle and posterior. These regions were differentiated according to the intensity of cytoplasmic protrusions. There was an increase in these projections in both larval populations in the Dowling and PI 227687 treatments, which contained the flavonoids genistein and rutin.

Among the treatments, the PI 227687 genotype was the one that provided more protrusions than the Dowling treatment, which could alter the insect's absorption and metabolism.

Among the larval populations, the resistant larvae had more protrusions than the susceptible larvae, which could be related to the insect's greater ability to degrade damaged cells.

REFERENCES

ABOT, A.R.; et al. Development of resistance by *Anticarsia gemmatalis* from Brazil and the United States to a nuclear polyhedrosis virus under laboratory selection pressure. **Biological Control**, v.7, n. 1, p.126-130, aug. 1996.

ANDERSON, E; HARVEY, W. Active transport by cercopia midgut. **The journal of Cell Biology**. V.31,n.1, p.107-134, Oct. 1966.

ARRUDA, W.; OLIVEIRA, G.M.; SILVA, I.G. Toxicity of the ethanolic extract of *Magonia pubescens* on *Aedes aegypti* larvae.**Revista da Sociedade Brasileira de Medicina Tropical**. V.36, n.1, p.17-25, jan-feb. 2003.

BROOKS, M.E. et al. Infection of its lepidopteran host by the *Helicoverpa armigera stunt virus* (Tetraviridae). **Journal of Invertebrate Pathology**, v. 80, p.97-111, 2002.

CHAPMAN, R.F. **The Insects: structure and function**. 4.ed. Cambridge: Harvard University Press, 1998.

CRISTOFOLOLETTI, P.T; RIBEIRO, A.F.; TERRA, W.T. Apocrine secretion of amylase, exocytosis of trypsin along the midgut of *Tenebrio molitor* larvae. **Journal Insect Physiology**, v.47, n.2, p.143-155, feb. 2001.

CRUZ, N.A. et al. Ultramorphology of the midgut cell surface of *Anicarsia gemmatalis* Hübner (Lepidoptera:Noctuidae) larvae resistant and susceptible to the nucleopolyhedrovirus *Ag* MNPV. **Neotropical Entomology** (In Press), 2010.

DE PRIESTER, W. Ultrastructure of the midgut epithelial cells in the fly *Callifora erythrocephala*. **Journal of Ultrastructure Research.** New York v.36, n.5, p. 783805, set. 1971.

FLIPSEN, J.T.M.; et al. Passage of *Autographa californica* nuclear polyhedrosis virus through the midgut epithelium of *Spodoptera exigua* larvae. **Virology**, New York v.208, p.328-335, 1995.

GAZZONI, D.L. HÜLSMEYER,A.; HOFFMANN-CAMPO, C.B. Effect of different doses of rutin and quercetin on the biology of *Anticarsia gemmatalis* Hübner, 1818 (Lep., Noctuidae). **Revista Agropecuaria Brasileira**, Brasilia, v.32, n.7, p. 673-681, jun./jul. 1997.

HARBONE, J.B. Flavonoid pigments. In: ROSENTHAL, G.A.; BEREMBAUM, M.R. (Eds) **Herbivores and their interaction with secondary metabolites**, London: Academic Press, 1991, p.389-429.

HOFFMANN-CAMPO, C.B; OLIVEIRA, E.B.; MOSCARDI,F. **Mass rearing of the soybean caterpillar.** Londrina: EMBRAPA, 1985.

HOFFMANN-CAMPO,C.B.; et al. Mechanisms of resistance of soybean genotypes: non-preference test for *Anticarsia gemmatalis* Hübner, 1818 (Lep.:Noctuidae).

Pesquisa Agropecuaria Brasileira, Brasilia, v.29, n.4, p. 513-519, abr. 1994.

HOFFMANN-CAMPO, C.B.; **Role of the flavonoids in the natural resistance of soyabean to *Heliothis virescens* (F.) and *Trichoplusia ni* (Hübner)**. 1995.

Dissertation (PhD in Botany) - 165 p. The University of Reading, Reading.

HOFFMANN-CAMPO, C.B.; et al. **Soybean pests in Brazil and their integrated management.** Londrina: EMBRAPA-CNPSo, 2000.

HOFFMANN-CAMPO, C.B.; et al. Pre-ingestive and and post-ingestive effects of soybean extracts and rutin on *Trichoplusia ni* growth. **Entomologia Experimentalis et Applicata**, Amsterdam, v.98, n.2, p.181-194, feb. 2001.

HOFFMANN-CAMPO, C.B.; et al. Detrimental effect of rutin on *Anticarsia gemmatalis*. **Pesquisa Agropecuària Brasileira**, Brasilia, v.41, n.10, p. 1453-1459, Oct. 2006.

JIMENEZ, D.R.; GILLIAN, M. Ultrastructure of the ventriculus of the honey bee, *Apis mellifera* (L.): cytochemical localization of acid phosphatase , alkaline phosphatase, and nonspecific esterase. **Cell and Tissue Research**, v.261, p.431-443, 1990.

KAKAMAND, F.A.K.H.; MAHMOUD,T.T.; AMIN,A.B.B. The role of three insecticides in disturbance the midgut tissue in honey bee *Apis mellifera* L. workers. **Journal Dohuk University**, Kurdistan v.11, n.1, p. 144-151, May. 2008.

KOGAN, M. Natural chemicals in plant resistance to insects. **Journal of Research**, Iowa State v. 60, p.501-527, 1986.

KUBO, I.; HANKE, F.G. Chemical methods for isolating and identifying phytochemicals biologically active in insects. In: MILLER, J.R.; MILLER, T.A. (Eds) **Insects plants interactions**. New York: Spring Verlog. 1986 p.225-249.

LEVY, S.M. *et al.* The larval midgut of *Anticarsia gemmatalis* (Hübner) (Lepidoptera: Noctuidae): light and electron microscopy studies of the epithelial cells. **Brazilian Journal of Biolpogy**, Sao Paulo, v.64, n. 3b, p.1-6, aug. 2004.

LEVY, S.M. **Susceptibility/Resistance of *Anticarsia gemmatalis* Hübner, 1818 (Lepidoptera:Noctuidae) larvae to nucleopolyhedrovirus (*AgMNPV*) infection: morphological study and detection of agglutinins in the midgut.**

2005.Thesis (Doctorate in Biological Sciences). Institute of Biosciences, Universidade Estadual Paulista-UNESP. Botucatu Campus, Botucatu.

LEVY, S.M.,et al.. Susceptibility/resistance of Anticarsia gemmatalis larvae to its nucleopolyhedrovirus (AgMNPV): structural study of the peritrophic membrane.

Journal of Invertebrate Pathology v.93, p.183-186, 2007.

LEVY, S.M. et al. Ultramorphology of digestive tract of *Anticarsia gemmatalis* (Hübner, 1818) (Lepidoptera:Noctuidae) at final larval development. **Semina Ciências Agrarias**, Londrina v. 29, n.2, p.313-322, Apr./Jun. 2008.

LEVY, S.M.;et al. A morphometric study of the midgut in resistant and non-resistant Anticarsia gemmatalis (Hübner) (Lepidoptera: Noctuidae) larvae to its nucleopolyhedrovirus (AgMNPV). **Journal of Invertebrate Pathology**, v.101, n.1, p.17-22, Apr. 2009.

MACHADO, V.O.F., et al. BIOLOGICAL ASPECTS OF *Anticarsia gemmatalis* HÜBNER (LEPIDOPTERA: NOCTUIDAE) IN SOYBEAN CULTIVARS (*Glycine max* MERRIL). **Pesquisa Agropecuària Tropical**, v. 29, n.1, p.39-41, 1999.

MENDONÇA E. G., et al. Determination of enzymatic activity and number of bacteria associated with the midgut of the soybean caterpillar *Anticarsia gemmatalis*, reared on different diets. **Revista Ceres**, v.56, n.1, p.18-24, jan./feb. 2009.

MOSCARDI, F.; CARVALHO, R.C.Z. Consumption and utilization of soybeans by *Anticarsia gemmatalis* Hüb. (Lepidoptera: Noctuidae) infected at different larval stages by its nuclear polyhedrosis virus. **Anais Sociedade Entomològica do Brasil**, Londrina, v.22, n.2, p.267-280, jun.1993.

PINHEIRO , D. O. **Regional variations in the epithelial cells of the midgut of *Diatrea saccharalis* Fabricius 1794 (Lepidoptera: Crambidae), in non-parasitized larvae and during parasitism by *Cotesia flavipes* (Hymenoptera: Braconidae).** 2006.Thesis (Doctorate in Biological Sciences: Zoology). Institute of Biosciences, Universidade Estadual Paulista-UNESP Botucatu Campus, Botucatu.

PINHEIRO, D.O. et al. Morphometric study of the midgut epithelium in the *Diatrea saccharalis* Fabricius (Lepidoptera: Pyralidae) larvae. **Neotropical Entomology**,Vacaria, v.32, n.3, p. 453-459, jul./set. 2003

PINHEIRO, D.O. et al. Morphometry of the midgut epithelium *of Diatrea saccharalis* Fabricius, 1974 (Lepidoptera) parasitized by *Cotesia flavipes* Cameron, 1891 (Hymenoptera). **Journal of Invertebrate Pathology**, v.93, n.1, p.60-62, set. 2006.

PINHEIRO, D.O. et al., Morphological Regional Differences of Epithelial Cells along the Midgut in Diatraea saccharalis Fabricius (Lepidoptera: Crambidae) Larvae.

Neotropical Entomology, Vacaria, v.37, n.4, p.413-419, jun./ago. 2008.

PIUBELLI, G.C et al. Are chemical compounds important for soybean resistance to *Anticarsia gemmatalis* ? **Journal of Chemical Ecology**, California, v.31, n.7, p.15091525, jul. 2005.

RIBEIRO A F, FERREIRA C, TERRA, W R Morphological basis of digestion in insects. **Acta Microscopica** v.6: p.71-73. 1997.

SANTOS, C.D. et al. The larval midgut of the cassava hornworm (*Erynnis ello*): ultrastructure, fluid and fluxes and the secretory activity in relation to the organization of digestion. **Cell and Tissue**

Research, v.237, p. 565-574, 1984.

TERRA, W.R.; FERREIRA, C. The digestion of insects. **Ciência Hoje,** v.12, p.28-38, 1991.

TERRA, W.R.; FERREIRA, C. Insect digestive enzymes: properties, compartmentalization and fuction. **Comparative Biochemistry and Physiology**, v. 109, p. 1-62, 1994.

WASHBURN, J.O., et al. Comparative pathogenesis of *Helicoverpa zea* M nucleopolyedrovirus in noctuidae larvae. **Journal of General Virology**, v.84, p.343351, 2001.

WOOD, A.R., LEHANE, M.J.Relative contributions of apocrine and eccrine secretion to digestive enzyme release from midgut cells of Stomoxys calcitrans (Insecta: Diptera). **Journal of Insect Physiology**, v. 37, p.161-166, 1991.

5 ARTICLE C - HISTOCHEMICAL STUDY OF THE MIDDLE INTESTINE OF *A. GEMMATALIS* LARVARS SUSCETIBLE AND RESISTANT TO THE AGMNPV VIRUS FEEDEDEDED WITH INSECT-RESISTANT SOYBEAN GENOTIES

SUMMARY

The midgut of insects is the main route of absorption and action of most chemical and biological insecticides and substances that promote plant resistance to insects (allelochemicals). The midgut may contain molecules that are extremely important for cell maintenance, as well as aiding in the metamorphosis of these insects. Among these molecules, polysaccharides, both neutral and acidic carbohydrates, proteins and lipids (among others) can be detected in histological preparations using special staining techniques, histochemistry, which identifies these molecules in the tissue. The aim of this study was to verify the chemical composition of the substances present in the midgut *of A. gemmatalis* larvae susceptible and resistant to the multiple nucleopolyhedrovirus (*Ag* MNPV) after treatment with leaves of soybean genotypes containing the flavonoid rutin (R) and the isoflavonoid genistein (G) in different concentrations. Both larval populations were treated with fresh soya genotypes containing rutin and genistein (BRS 257, control; BR 16, G=0.0016mg/g; Dowling, G= 0.0429; PI 229358, R=0.0212 and G= 0.0136mg/g; IAC 100 R=0.0972 and G= 0.0142 mg/g and PI 227687, R=0.3682 and G= 0.0122mg/g). The midguts of the fourth instar larvae were fixed and processed for the detection of neutral carbohydrates, acids, proteins and lipids. PAS revealed that resistant larvae treated with the IAC 100 and PI 227687 genotypes had a higher concentration of neutral polysaccharides and after diastasis the IAC 100 treatment in resistant larvae had the highest concentration compared to the control and the other treatments. In the detection of total proteins, the susceptible larvae showed a positive reaction in all treatments for the three techniques applied, with intense nuclear marking, while the resistant larvae showed a positive reaction in treatments IAC 100 and PI 227687 (bromophenol mercuric blue); IAC 100 (bromophenol blue). The lipid detection reaction was only positive in resistant larvae for treatments PI229358, IAC 100 and PI227687. In the detection of acid carbohydrates, there was no positive reaction for any of the treated genotypes, in both larval types.

Keyword: Anticarsia gemmatalis. Rutin. Genistein. Carbohydrates. Proteins. Lipids.

1 INTRODUCTION

In insects, the feeding mechanism, its structure and the digestive chemistry are processed in the digestive system, which is made up of the alimentary canal or digestive tube and attached organs. The digestive tract is made up of three distinct morphofunctional regions: the anterior intestine, the middle intestine and the posterior intestine (CHAPMAN, 1998), with the middle intestine (MI) being the main site of digestion, food absorption and the main access route for the chemical and biological insecticides administered (MOSCARDI; CARVALHO, 1993).

Currently, the fight against pests in general tends to gradually reduce the use of agrochemicals due to their side effects, being replaced or associated with alternative methods of control that take into account economic and ecological aspects, for the medium and long term solution. The strategy is a proposal for coexistence with pests, giving biological control a chance and recommending chemical control when the pest population reaches levels that cause greater damage than the costs of control (HOFFMANN-CAMPO et al., 2006). This proposal is based on ecological, economic and social knowledge, aiming to interfere as little as possible in the agro-ecosystem and is known as Integrated Pest Management (IPM), and can be characterized by the alignment of control methods with ecological, economic and social principles, aiming to interfere as little as possible in the agro-ecosystem (GAZZONI et al., 1988; PANIZZI, 1990).

The damage caused by a key soybean pest, the soybean caterpillar *Anticarsia gemmatalis*, has been minimized by biological control using the nucleopolyhedrovirus (*Ag* MNPV). The viral polyhedra, when ingested by the caterpillar, reach the lumen of the IM, where their protein capsule is solubilized by the action of the alkaline environment of this region, releasing the nucleocapsids (GRANADOS, 1981). Studies show that the epithelium of the IM is the first tissue to be infected by the virus during the infection process (KAWANISHI et al., 1972). For this infection process to take place, the nucleocapsids need to cross the PM, the first barrier against pathogens; the nucleocapsids penetrate through its pores, reaching the inside of the cell after fusing with the membrane of its microvilli, replicating in the nuclei of the columnar cells and transmitting the infection to other tissues (FLIPSEN et al., 1995; VOLKMAN, 1997).

The efficiency of the virus in biological control and the low cost of production have increased its use, which may lead to greater selection pressure on the insect population in the field. Laboratory experiments involving selection pressure with AgMNPV resulted in rapid selection for resistance to this entomopathogen (ABOT et al., 1996). The appearance of larvae resistant to the nucleopolyhedrovirus, even under laboratory conditions, has led researchers to question which mechanisms are involved in this resistance. Some studies indicate that this resistance could be attributed to differential defense mechanisms in the IM (ABOT et al., 1996).

Soybean resistance to insects is a plant defense mechanism that encompasses a series of morphological characteristics and also a complex of chemical substances that can make it repellent, toxic or otherwise unsuitable for insect pests (PIUBELLI, 2004). In soybeans, the most likely defense substances are flavonoids (HOFFMANN-CAMPO, 1995), which are generally more abundant in pest-resistant genotypes. Flavonoids are a group of natural substances that have a wide range of biological activities. One of the major actions of flavonoids is related to the formation of fatty acids by the action of phospholipase A2, responsible for the hydrolysis of phospholipids present in cell membranes, with the release of arachidonic acid (KUBO; HANKE, 1986; LARA, 1991; SILVA et al., 2002; ONYILAGHA; GROTEWOLD, 2004; LOPES et al., 2005). Thus, flavonoids can interact and penetrate the lipid bilayers of membranes, altering the barriers and permeability of cell membranes

(LIMA et al., 1999; SILVA et al., 2001). Other functions could be related to acting on the expression of membrane receptors, as antibacterial and antiviral agents and inhibitors of proteolytic enzymes (TOLEDO et al., 2003).

The phenolic compounds present in plants have different effects on herbivores. The activation of these compounds occurs through oxidation and can lead to the precipitation of proteins (APPEL, 1993). The hydroxyl radicals formed during the oxidation of phenolic compounds have a toxic action and are responsible for disrupting membrane integrity and metabolic disorders in the intestinal epithelium (APPEL, 1993). In caterpillars, the covalent bond between chlorogenic acid and monomeric proteins inhibits the absorption of amino acids (FELTON et al., 1989; FELTON; DUFFEY, 1991).

Seeking to understand the effects of flavonoids on insects, HOFFMANN-CAMPO et al. (1994) carried out food preference tests on *A. gemmatalis* and noted that the soybean genotypes BR82-12547, IAC74-2832, PI 227687, PI 229358, PI 274454 were the most rejected by the caterpillar. Later studies identified the presence of seven flavonoids in PI 227687, one of the most widely used in insect resistance programs, which could indicate greater toxicity of this cultivar and explain the insect's rejection. Among the flavonoids, rutin (quercitin 3-O-rutinoside) is recognized for playing a very important role in plant defense against lepidopterans (HOFFMANN-CAMPO, 1995; HOFFMANN-CAMPO et al, 2001). This substance has also been identified in extracts from the leaves of PI 227687, PI 274454 and a few other insect-resistant genotypes (PIUBELLI et al, 2005). Previous studies of these substances on defoliator insects have shown antibiotic and/or non-consumptive effects on Lepidoptera larvae such as *Manduca sexta* (L.) (STAMP; SKROBOLA, 1993), *Heliothis virescens* (F.) (HOFFMANN-CAMPO, 1995) and *Trichoplusia ni* (Hübner) (HOFFMANN-CAMPO et al., 2001). In *A. gemmatalis,* rutin showed characteristic antibiosis effects, lengthening the larval cycle and reducing survival (GAZZONI et al., 1997). In addition, the isoflavonoid genistein (7-O-β-D-glucoside) was observed in lower concentrations in most of the leaf extracts of the genotypes studied by Piubelli et al. (2005) and used by breeding programs as a source of resistance to defoliator insects.

Isoflavonoids are a distinct class of flavonoids and demonstrate a wide range of biological properties, but probably the most important are the estrogenic, antifungal and bactericidal activities of phytoalexins and the insecticidal activity of rotenoids (HARBORNE, 1991). However, little is known about their effects on food preference and pest biology, as well as the action of a plant defense chemical on the biological activity of the insect (HOFFMANN-CAMPO et al., 2001).

Both biological and plant resistance controls against insects use the digestive system, especially the IM, for their action, since they need to be ingested by the caterpillars. However, there are few studies on the nutritional and post-ingestive effect of flavonoids on the epithelial cells of the MI, whether or not they are infected with *Ag* MNPV. Some studies have shown that rutin has increased the mortality of resistant *A. gemmatalis* larvae compared to susceptible larvae (PIUBELLI et al., 2006). The lack of knowledge about how rutin acts, as well as the mechanisms of action of this substance in reducing resistance in *A. gemmatalis* larvae, reinforces the need for further studies on the MI of these insects, a possible site of action of flavonoids and the main route of access for *Ag* MNPV.

In *A. gemmatalis* larvae, the IM is formed by a pseudostratified columnar epithelium, with four distinct cell types: columnar, goblet, regenerative and endocrine cells (LEVY et al., 2004), as described in other Lepidoptera species (CHIANG et al., 1986; BINDER; BOWERS, 1994; BILLINGSLEY; LEHANE, 1996). The epithelium is lined by an acellular structure called the peritrophic membrane (PM) which separates the epithelium from the intestinal contents, protecting it from abrasion, acting as a permeable barrier to digestive enzymes and digestive products (EISEMANN; BINNINGTON, 1994; TERRA, 2001), as well as forming a barrier that hinders or prevents microorganisms from entering the body cavity (WANG; GRANADOS, 2001). Supporting the tube and separating it from the hemocele are two layers of muscle, the innermost, the circular muscle, and the outermost, the longitudinal muscle.

Morphological studies of the midgut in *A. gemmatalis* larvae showed structural differences in this organ between susceptible and resistant larvae. In the latter, the epithelial layer was higher, with a greater number of regenerative cells; the columnar cells had a greater number of cytoplasmic protrusions compared to susceptible larvae (LEVY et al., 2005).

It is known that these different cell types are involved in different processes and functional activities. Columnar cells participate in the process of secretion of digestive enzymes, peritrophic membrane and absorption of assimilated products of digestion (RICHARDS; DAVIES, 1994; LEHANE; BILLINGSLEY, 1996); goblet cells are related to absorption of metabolites and ionic homeostasis (MOFFET et al., 1995; BILLINGSLEY; LEHANE, 1996; CHAPMANN, 1998); regenerative cells play an important role in epithelial renewal (LEHANE; BILLINGSLEY, 1996; CHAPMANN, 1998; CAVALCANTE; CRUZ-LANDIM, 1999) and endocrine cells play a role in hormonal control, digestion, diuresis, development, reproduction and peristatic movements (BROWN et al., 1985; PRATT et al., 1989; WOODHEAD et al., 1989; SCOOFS et al., 1993, COAST; KAY, 1994; PATEL et al., 1995).

Due to the functional variation of the IM, cells present substances that help to maintain the organ and the cell itself, which may be involved in the production of energy (carbohydrates), energy reserves (lipids) and protein synthesis that can be identified using histochemical techniques. However, little is known about the chemical composition of the GI of *A. gemmatalis* larvae, and there is no information on the presence of these substances in larvae ingesting resistant plants. The histochemical study of the components of the midgut is therefore necessary in order to verify whether flavonoids can alter the normal activity of the MI in the two larval populations of *A. gemmatalis* and correlate the findings with the mechanisms of resistance to *AgMNPV*.

2 .MATERIAL AND METHODS

The following soybean genotypes were used: BRS 257 as a control, the BR 16 genotype with insect susceptibility characteristics according to Salvador (2008) and the genotypes with insect resistance characteristics: Dowling (0.0 - 0.0429 mg/g of rutin and genistein); PI 229358 (0.0212 - 0.0136 mg/g of rutin and genistein); IAC 100 (0.0972 - 0.0142 mg/g of rutin and genistein) and PI 227687 (0.3682 - 0.0122 mg/g of rutin and genistein), from Embrapa Soja Londrina-PR. Larvae of *A. gemmatalis* larvae susceptible and resistant to *AgMNPV*, from eggs selected by the entomology laboratory (EMBRAPA- CNPSoja) were fed the leaves of these genotypes until they reached the fourth instar (10 - 12 days post-hatch); They were anaesthetized by cooling, dissected under a stereoscope in a saline solution for insects (1.80 g NaCl; 1.88 g KCl; 0.16 CaCl; 0.004 g $NaHCO_3$ and distilled water- q.s.p 100 mL), the ends were attached by entomological pins, and with the help of ophthalmic scissors a logitudinal incision was made in the ventral region, from the first thoracic segment to the penultimate abdominal segment. The cuticle was folded to one side, 30 tubes were collected per treatment and fixed for 6 hours in Karnovsky solution according to Cerri and Cerri-Sasso (2003), for the detection of Neutral Polysaccharides (PAS), glycogen, acid polysaccharides and total proteins. For lipids, the protocol of HERNANDEZ-BLAZQUEZ et al (1989) was used in a fixing solution of 4% paraformaldehyde, 1% calcium chloride and 1% cadmium chloride for 24 hours after being fixed in osmium. The materials were then dehydrated in an increasing series of ethyl alcohol and embedded in historesin (Historesin®, Leica embedding kit). Sections measuring 4µm were obtained with a tungsten razor on a LEICA RM 2265 automatic rotary microtome and submitted to the proposed techniques.

2.1. detection of neutral polysaccharides

To detect neutral polysaccharides, the sections were treated with 0.1% periodic acid at 56°C in an oven (15 minutes) and Schiff's Reagent (60 minutes) (MACMANUS - HOTCHKISS, in LISON, 1950), controlled by salivary amylase (to exclude the presence of glycogen) and blocked by acetylation according to McManus and Cason (to check for vic-glycol groups), counterstained with Harris's Hematoxylin (35 minutes), washed, dehydrated in an increasing series of ethyl alcohol (70 to 100%), diaphanized in xylene, mounted with Canada balsam and analyzed under the Zeiss Axiphot light photomicroscope of the Histology Department at UEL.

2.2. GLYCOGEN DETECTION

To detect glycogen, the slides were previously treated with 0.5% amylase for 2 hours in an oven at 37°C (FIORE-DONNO; BAUME, 1966), subjected to PAS and analyzed.

2.3. DETECTION OF ACID PoLISSACARIDES

For this technique, Toluidine Blue solutions were used for 15 minutes at pH 1.0 (highly sulphated acidic polysaccharides); pH 2.5: (sulphated acidic polysaccharides); pH 4.0: (acidic polysaccharides), according to the PEARSE protocol (1972).

2.4. total protein determination

2.4.1 Fast Green 0.1% method According to the Mello and Vidal Protocol (1980)

Treatment of the sections with 0.1% Fast Green solution in 1% acetic acid pH 2.7 (30 minutes), dehydration in butanol, diaphanization in xylene, mounting and analysis.

2.4.2 Bromophenol Mercuric Blue Method According to Pearse (1972)

Treatment of the sections in 0.05% Bromophenol Blue solution and 1% Mercuric Chloride in 2% acetic acid (2 hours); washing in 0.5% aqueous acetic acid solution (5 minutes), differentiation in butyl alcohol, diaphanization, mounting and analysis.

2.4.3 Bromophenol Blue According to Coello Protocol (1989)

Treatment of the sections in 0.1% Bromophenol Blue solution in 95% ethyl alcohol (15 minutes), washing of the sections in 0.5% aqueous acetic acid solution (5 minutes), rapid diaphanization in Xylene and mounting with Canada balsam.

2.5. LIPIDEOS DETECTION

For this technique, the digestive tubes were fixed according to the protocol of Hernàndez-Blazquez et al (1989) in a fixative solution of 4% paraformaldehyde, 1% calcium chloride and 1% cadmium chloride for 24 hours, post-fixed in 1% osmium tetroxide (0.1M phosphate buffer pH 7.2) for 3 hours, dehydrated in ethyl alcohol (90%, 100%), embedded in historesin (Historesin®, Leica embedding kit), cut at 4μm with a tungsten knife, on a LEICA RM 2265 automatic rotary microtome, placed on glass slides previously cleaned with alcohol ether.

After obtaining the histological sections, they were treated with hydrogen peroxide solution (10 minutes), stained with 1% Sudan Black B solution in 70% ethyl alcohol (15 minutes), differentiated in 70% ethyl alcohol, washed, mounted with glycerin and analyzed immediately.

3 RESULTS

3.1 .NEUTRAL POLYSACCHARIDES

The results of the histochemical detection of neutral carbohydrates using the Periodic Acid-Schiff (PAS) method are shown in Table 3.1.

Table 3.1 - Results of the PAS reaction to detect neutral polysaccharides in the midgut of *A. gemmatalis* larvae susceptible (LS) and resistant (LR) to *AgMNPV*.

		Genotypes											
		BRS 257 R: 0,0 G:0,0		BR 16 R: 0,0 G:0,0016		Dowling R: 0,0 G:0,0136		PI 229358 R: 0.0212 G: 0,0136		IAC 100 R:0,0972 G: 0,0142		PI 227687 R: 0.3682 G:0.0122	
Structures		LS	LR	LS	LR	LS	LR	LS	LR	LS	LR	LS	LR
Peritrophic membrane		-	-	-	-	-	-	-	-	-	-	-	-
	Grooved edge	-	-	-	-	-	-	-	-	-	-	-	-
Cells	Apical portion	+++	++++	-	+	+	+	++	-	++	+++	++	+++
Columnar	Average portion	+++	++++	+	+	++	+	++	-	++	+++	-	+++
	Basal Portion	+++	++++	+	+	++	-	-	-	++	+++	-	+++
	Protrusions	-	-	++	-	-	-	+++	-	-	-	-	+++

Goblet cells	-	+++	-	-	-	-	++	-	-	-	-	-
Regenerative cells	-	-	-	-	-	-	+	-	-	-	-	-
Muscle Layer	-	-	-	-	-	-	-	-	-	-	-	-

Legend: (-) negative reaction; (+) weak reaction; (++) moderate reaction; (+++) strong reaction; (++++) very strong reaction.

The midgut of *A. gemmatalis* larvae susceptible and resistant to *AgMNPV* treated with leaves of the control genotype (BRS 257) showed a positive PAS reaction only in the epithelial layer, with the columnar cells being responsible for the positivity in both larval types. The reaction was seen in the three apical, middle and basal portions (Figs. 4.1 A and 4.2 A).

In general, the reaction was more intense in the epithelial cells of the resistant larvae. In addition, the LR larvae showed a strong reaction in the goblet cells (Fig. 4.2 A).

2.5.1 Larvae Susceptible to MNPV *Ag*

In the treatment with the BR 16 genotype, the columnars showed weak positivity in the middle and basal portions, while the protrusions showed a moderately positive reaction. The other structures were not positive to the technique (Fig. 4.1 B).

When the Dowling genotype was treated, the columnar cells showed moderately positive granules in the middle and basal portions, while the apical portion showed a weakly positive reaction to PAS (Fig. 4.1 C).

For the PI 229358 genotype, positivity was observed in the protrusions of columnar cells, goblet cells and regenerative cells, with the intensity of staining varying between them. The columnar cells showed PAS-positive granules with a moderate reaction in the apical and middle portions, while in the cytoplasmic protrusions, the staining was intense (Fig. 4.1 D). The goblet cells showed moderate positivity and the regenerative cells showed a weak reaction to PAS.

In the treatment with the IAC 100 genotype, moderate positivity to the method was observed in the three cytoplasmic portions of the columnar cells (Fig. 4.1 E). The other structures showed no reactivity to the method.

For the PI 227687 genotype, only the apical portion of the columnar cells showed slight granulation with moderate positivity (Fig. 4.1F), and the other structures showed no positivity to the method.

2.5.2 Larvae Resistant to AgMNPV

The PAS reaction was weakly positive in the LR treated with the BR16 genotype, with the presence of discrete granules in the apical, middle and basal portions of some columnar cells (Fig. 4.2 B). The other structures were not positive to the method.

In the treatment with the Dowling genotype, only a few columnar cells reacted weakly and showed positive granulation in the apical and middle portions, sometimes being more concentrated in the apical portion (Fig.4.2 C). The other structures were PAS negative.

For the PI 229358 genotype, there was a weak positive reaction in one cell to PAS (Fig. 4.2 D).

In the treatment with the IAC 100 genotype, positive granules were observed in the three portions of the columnar cells (apical, middle and basal) which provided a strong reaction intensity. The other structures were negative to the method (Fig. 4.2 E)

For the PI 227687 genotype, columnar cells and protrusions showed strong positivity to the method (Fig. 4.2 F). The other structures did not react to the method.

3.2. Glycogen

The results of the histochemical detection of glycogen, preceded by diastase for larvae susceptible and resistant to MNPV *Ag*, are shown in Table 3.2.

Table 3.2 - Histochemical results for glycogen detection. PAS reaction preceded by salivary amylase, in the midgut of *A. gemmatalis* larvae susceptible (LS) and resistant (LR) to *AgMNPV*.

		Gen				**lotuses**							
		BRS 257 R: 0 G:0		BR 16 R: 0 G:0,0016		Dowling A: 0 G:0,0136		PI229358 R: 0.0212 G: 0,0136		IAC 100 R:0,0972 G: ,0142		PI227687 R: 0,3682 G:0,0122	
Structures		LS	LR	LS	LR	LS	LR	LS	LR	LS	LR	LS	LR
Peritrophic membrane		-	-	-	-	-	-	-	-	-	-	-	-
	Grooved edge	-	-	-	-	-	-	-	-	-	-	-	-
Cells	Apical portion	+++	+	-	-	+	-	++	-	+	++	++	+++
Columnar	Average portion	+++	+	+	-	++	+	++	-	+	++	-	+++
	Basal Portion	+++	+	+	-	++	-	-	-	+	++	-	+++
	Protrusions	-	-	+	-	-	-	+++	-	-	-	-	+++
Goblet cells		-	++	-	-	-	-	++	-	-	-	-	-
Regenerative cells		-	-	-	-	-	-	+	-	-	-	-	-
Muscle Layer		-	-	-	-	-	-	-	-	-	-	-	-

Legend: (-) negative reaction; (+) weak reaction; (++) moderate reaction; (+++) strong reaction.

After diastase, it was observed that for the control treatment there was a large reduction in positivity in the columnar and goblet cells of the resistant larvae, while the susceptible larvae showed no change in reaction positivity.

For treatment with the BR 16 genotype, the midgut showed a reduction in positivity in all the structures marked in the IM of the LR, while in the LS there was a reduction in positivity only in the protrusions of the columnar cells.

In the Dowling treatment, the resistant larvae showed a reduction in positivity only in the apical portion of the columnar cells and a maintenance of intensity in the other portions, while the susceptible larvae showed no change in the MI structures.

In the IAC 100 treatment there was a similar reduction in both populations, both in susceptible and resistant larvae.

The IM of susceptible and resistant larvae fed PI 229358 and PI 227687 showed no change in reaction positivity.

3.3. polysaccharide acids

The detection of polysaccharides in the midgut of *A. gemmatalis* larvae susceptible and resistant to AgMNPV by the Toluidine Blue reaction at pH 1.0, 2.5 and 4.0 did not show a positive reaction in all the methods tested.

3.4. Total Proteins

3.4.1 Fast Green 0.1%

The results of total protein detection using the Fast Green 0.1% method are shown in Table 3.3.

The midgut of LS and LR fed on leaves of the control genotype BRS 257 was positive for the method in the goblet cells and muscle layer. There was a difference in the intensity of the reaction, with LS showing a moderate reaction while resistant larvae showed a weak reaction (Figs. 4.3 A and G).

In LS, the other structures did not show a positive reaction to the method applied, but in resistant larvae, a positive reaction could be seen in the columnar and regenerative cells, which had a weak reaction to the method (Fig. 4.3 G).

Table 3.3 - Histochemical results for the detection of total proteins in the midgut *of A.gemmatalis* larvae susceptible (LS) and resistant (LR) to *AgMNPV* using the Fast Green 0.1% method.

Structures	Genotypes											
	BRS 257 R: 0,0 G: 0,0		BR 16 R: 0,0 G: 0,0016		DOWLING R: 0,0 G: 0,0429		PI 229358 R: 0,0212 G: 0,0136		IAC 100 R: 0,0972 G: 0,0142		PI 227687 R: 0,3682 G: 0,0122	
	LS	LR	LS	LR	LS	LR	LS	LR	LS	LR	LS	LR
Membrane Peritrophic	-	-	-	-	-	-	-	-	-	-	-	-
Cells Columnar	-	+	+	-	+	-	++	-	+	-	++	-
Cells Caliciforms	++	++	+	-	++	-	++	-	++	-	++	+
Cells Regenerative	-	+	-	-	-	-	-	-	-	-	-	-
Layer Muscular	++	+	-	-	+	-	-	-	-	-	++	-

Legend: (-) negative reaction; (+) weak reaction; (++) moderate reaction.

3.4.1.1 Larvae susceptible to *Ag* MNPV

In the treatment with the BR 16 genotype there was a weak reaction in the columnar and goblet cells, the other structures were not positive (Fig. 4.3 B).

For the Dowling genotype, the goblet cells showed a moderate reaction, the columnar cells and the muscle layer showed a weakly positive reaction (Fig.4.3 C), while the other structures were not positive.

In the PI 229358 treatment, the columnar and goblet cells showed a moderate reaction and the other structures did not show a positive reaction (Fig. 4.3 D).

The IM of the LS treated with the IAC 100 genotype showed a positive reaction in the columnar and goblet cells, varying only in the intensity of the reaction, in the columnar cells the reaction was weak while in the goblet cells the reaction was moderate. The other structures showed no reaction to this method (Fig. 4.3 E).

In the larvae treated with the PI 227687 genotype, a moderately positive reaction was observed in the columnar and goblet cells and in the muscle layer (Fig. 4.3 F). The other structures showed no reaction to the method applied.

3.4.1.2 Larvae resistant to AgMNPV

Only the goblet cells of the IM epithelium of the larvae fed the PI 227687 genotype showed a weak reaction. The other treatments were not positive.

3.4.2 Bromophenol blue Mercury

In general, all the midgut structures of the two larval types reacted positively to the method applied. Table 3.4 shows the reaction intensity values of the susceptible larvae.

Table 3.4 - Histochemical results for total protein detection in the midgut of *A. gemmatalis* larvae susceptible (LS) to AgMNPV - Bromophenol mercuric blue method according to Pearse (1972).

Structures		Genotypes					
		BRS 257 R: 0,0 G: 0,0	BR 16 R: 0,0 G:0,016	Dowling R: 0,0 G: 0,0429	PI 229358 R: 0,0212 G: 0,0136	IAC 100 R: 0,0972 G: 0,0142	PI 227687 R: 0,3682 G: 0,0122
Peritrophic membrane		-	-	-	-	-	-
	Protrusions	+	+++	+	+++	++	-
Columnar cells	Cytoplasm	++	++	++	++	+++	+++
	Core	++++	++++	++++	++++	++++	++++
Goblet cells	Cytoplasm	+++	+++	+++	+++	+++	+++
	Core	++++	++++	++++	++++	++++	++++
Regenerative cells	Cytoplasm	-	++	+	-	++	+
	Core	++++	++++	++++	++++	++++	+++
Muscle Layer		+	+	+	+	+	+

Legend: (-) negative reaction; (+) weak reaction; (++) moderate reaction; (+++) strong reaction; (++++) very strong reaction.

AgMNPV-susceptible and -resistant larvae fed the different genotypes showed the same very strong pattern of nuclear staining in all the epithelial cells.

In the control treatment (BRS 257), the goblet cells showed a strong cytoplasmic reaction for both larval types (Figs. 4.4 A and 4.5 A), while the columnar cells showed a moderate reaction intensity (Figs. 4.4 A and 4.5 A). In addition, the regenerative cells showed moderate positivity in the LR, unlike the LS, which showed no positivity.

The striated border, protrusions and muscular layer in both larval types showed weak positivity, while the peritrophic membrane showed no reaction.

3.4.2.1 Larvae susceptible to *Ag* MNPV

In the treatment with the BR 16 genotype, the regenerative and columnar cells showed moderate reaction throughout the cytoplasm (Fig. 4.4 B), but it was observed that the columnar protrusions were well marked with variations in the intensity of staining quite evident in those that had nuclear material inside (Fig. 4.4 B), while the goblet cells showed strong cytoplasmic reaction (Fig. 4.4 B). The MP was not positive for the method and the muscle layer showed a weak reaction in spots (Fig. 4.4 B). Some structures were visualized with variations from dark blue to red, such as the three cell types and the protrusions. This may be related to the concentration of proteins found in these midgut structures.

Positivity to the method was observed in all structures for the Dowling treatment, with the exception of the peritrophic membrane. The cytoplasm of the goblet cells showed a strong reaction; the columnar cells showed a moderate reaction and the regenerative cells showed a weak reaction (Fig. 4.4 C). The muscle layer showed a weak reaction. This treatment also showed a variation in the color tone of the reaction between blue and red (Fig. 4.4 C).

In the PI 229358 treatment, positivity was observed in the cytoplasm of the columnar and goblet cells and in the muscle layer (Fig. 4.4 D). Regenerative cells and the peritrophic membrane did not show positivity to the method.

The goblet cells showed a strong reaction, while the columnar cells showed a moderate cytoplasmic reaction and a strong reaction in the protrusions (Fig. 4.4 D). The muscle layer showed little reaction to the technique applied.

In this treatment, the reaction color shades can vary from blue to red.

In the treatment with the IAC 100 genotype, the goblet and columnar cells showed a strong cytoplasmic reaction, but the protrusions showed moderate reactivity, similar to that shown by the regenerative cells (Fig. 4.4 E), while the muscle layer showed weak positivity in small fragments of muscle (Fig. 4.4 E). The peritrophic membrane showed no positivity.

For the PI 227687 genotype, the cytoplasm of the columnar and goblet cells showed strong positivity, while the regenerative cells and the muscle layer showed weak reactivity (Fig. 4.4 F). No reactivity was observed in the peritrophic membrane.

3.4.2.2 Larvae resistant to AgMNPV

Table 3.5 shows the reaction intensity values of the resistant larvae.

For the columnar cells, the nuclear reactions were: very strong (BRS 257, IAC 100 and PI 227687) and negative for the other genotypes (Figs. 4.5 A, B and C). Cytoplasmic reactions were moderate (BRS 257, PI 227687), weak for IAC 100 and negative for the other genotypes (Fig. 4.5 B).

For the goblet cells, the nuclear reactions were: very strong BRS 257, IAC 100 and PI 227687 (Figs. 4.5 A, B and C), moderate Dowling and negative BR16 and PI229358. The cytoplasm of these cells showed a reaction: strong IAC 100, PI 227687 (Figs. 4.5 B and C), moderate BRS 257 and Dowling and negative for the others (Fig. 4.5 A).

The regenerative cells showed very strong nuclear reactions for the BRS 257 IAC 100 and PI 227687 genotypes (Figs. 4.5 A, B and C) and negative reactions for the other genotypes. The cytoplasm showed a weak reaction for PI 227687 and negative reactions for the other genotypes (Fig. 4.5 C).

The muscle layer showed a weak reaction for the BRS 257 genotype (Fig. 4.5 A) and a negative reaction for the other genotypes, while the peritrophic membrane was negative in all treatments.

Table 3.5 - Histochemical results for the detection of total proteins in the midgut *of A. gemmatalis* larvae resistant (LR) to AgMNPV - Bromophenol mercuric blue method according to Pearse (1972).

Structures		**Genotypes**					
		BRS 257 R: 0,0 G: 0,0	BR 16 R: 0,0 G: 0,016	Dowling R: 0,0 G: 0,0429	PI 229358 R: 0,0212 G: 0,0136	IAC 100 R: 0,0972 G: 0,0142	PI 227687 R: 0,3682 G: 0,0122
Peritrophic membrane		-	-	-	-	-	-
	Protrusions	++	-	-	-	-	-
Cells	Cytoplasm	++	-	-	-	+	++
Columnar	Core	++++	-	-	-	++++	++++
Cells	Cytoplasm	++	-	++	-	+++	+++
Caliciforms	Core	++++	-	++	-	++++	++++
Cells	Cytoplasm	-	-	-	-	-	+
Regenerative	Core	++++	-	-	-	++++	++++
Muscle Layer		+	-	-	-	-	-

Legend: (-) negative reaction; (+) weak reaction; (++) moderate reaction; (+++) strong reaction; (++++) very strong reaction.

3.4.3 Bromophenol blue

The results of total protein detection using the bromophenol blue method according to Coello 1989 are shown in Table 3.6.

Table 3.6 - Histochemical results for total protein detection in the midgut of *A. gemmatalis* larvae susceptible (LS) to AgMNPV - Bromophenol blue method according to Coello (1989).

Structures		**Genotypes**					
		BRS 257 R: 0,0 G: 0,0	BR 16 R: 0,0 G: 0,016	Dowling R: 0,0 G: 0,0429	PI 229358 R: 0,0212 G: 0,0136	IAC 100 R: 0,0972 G: 0,0142	PI 227687 R: 0,3682 G: 0,0122

Peritrophic membrane		-	-	-	-	-	-
Cells	Protrusions	-	+++	-	++	-	-
Columnar	Cytoplasm	++	+	++	+	+	+
	Core	++++	-	++++	+++	+++	+++
Cells	Cytoplasm	+++	-	+++	++	++	+
Caliciforms	Core	++++	-	++++	+++	+++	+++
Cells	Cytoplasm	-	-	-	-	-	+
Regenerative	Core	++++	+	+++	+	+++	+++
Muscle Layer		++	++	-	+	-	-

Legend: (-) negative reaction; (+) weak reaction; (++) moderate reaction; (+++) strong reaction; (++++) very strong reaction.

3.4.3.1 Larvae susceptible to *Ag* MNPV

This methodology showed a very strong reaction in the nuclei of the columnar, goblet and regenerative cells in the susceptible larvae treated with the BRS 257; IAC 100; PI 227687 genotypes (Figs. 4.6 A, E and F), while the larvae treated with the Dowling genotype showed a very strong reaction in the nuclei of the columnar and goblet cells and a strong reaction in the nucleus of the regenerative cells (Fig. 4.6 C).

The larvae treated with the PI 229358 genotype showed a strong reaction in the nuclei of the columnar and goblet cells and a weak reaction in the nucleus of the regenerative cells (Fig. 4.6 D).

As for the larvae treated with the BR 16 genotype, only a weakly positive result was observed in the nucleus of the regenerating cells (Fig. 4.6 B).

3.4.3.2 Larvae resistant to AgMNPV

The results of the detection of total proteins in *AgMNPV-resistant* larvae using the bromophenol blue method according to Coello 1989 are shown in Table 3.7.

Table 3.7 - Histochemical results for the detection of total proteins in the midgut *of A. gemmatalis* larvae resistant (LR) to *AgMNPV* - Bromophenol blue method according to Coello (1989).

Structures		**Genotypes**					
		BRS 257 **R: 0,0** **G: 0,0**	**BR 16** **R: 0,0** **G: 0,016**	**Dowling** **R: 0,0** **G: 0,0429**	**PI 229358** **R: 0,0212** **G: 0,0136**	**IAC 100** **R: 0,0972** **G: 0,0142**	**PI 227687** **R: 0,3682** **G: 0,0122**
Peritrophic membrane		-	-	-	-	-	-
	Protrusions	++	-	-	-	-	-
Cells	Cytoplasm	+	-	+	-	+	++
Columnar	Core	+++	-	+++	-	+++	++++
Cells	Cytoplasm	++	-	-	+	++	-
Caliciforms	Core	+++	-	-	++	+++	-
Cells	Cytoplasm	-	-	-	-	+	-
Regenerative	Core	+	-	-	-	+++	+
Muscle Layer		+	-	-	-	-	-

Legend: (-) negative reaction; (+) weak reaction; (++) moderate reaction; (+++) strong reaction; (++++) very strong reaction.

The columnar cells showed nuclear reactions: very strong (PI227687); strong (BRS 257, Dowling and IAC100), negative (BR 16 and PI 229358). The cytoplasmic reactions were: moderate (PI 227687); weak (BRS 257, Dowling and IAC 100) and negative BR16 and PI 229358 (Figs. 4.6 G, H and I).

The goblet cell nuclei showed reactions: strong in the BRS 257 and IAC 100 treatments (Figs. 4.6 G and H); moderate (PI229358) and negative (BRI6, Dowling and PI 227687). The cytoplasmic reactions were: moderate in the BRS 257 and IAC 100 genotypes (Figs. 4.6 G and H), weak for the PI 229358 treatments and negative in the BR16, Dowling and PI 227687 genotypes.

For regenerative cells, the nuclear reactions were: strong in the IAC 100 treatment (Fig. 4.6 H), weak in the BRS 257 and PI 227687 genotypes (Figs. 4.6G and I) and negative (BR16; Dowling, PI 229358), while the cytoplasm showed a weak reaction for the IAC 100 genotype and negative for the other genotypes (Fig. 4.6 H).

3.5. LIPiDEOS

After the histochemical reaction for lipid detection, the midgut of susceptible *A. gemmatalis* larvae did not show positive lipid granules in the midgut epithelium of these larvae; however, in resistant larvae there were positive scores in some treatments.

In resistant larvae treated with the IAC 100 genotype, a reaction can be seen in the columnar and goblet cells, indicating the presence of structural lipids (membrane formers) and some positive granules in the cytoplasmic protrusions (Fig. 4.7 B).

In the PI 229358 treatment, the positive granules can be seen in the columnar cells a , the other structures had no reaction to the method (Fig. 4.7 A).

For the resistant larvae from the PI 227687 treatment, the midgut epithelium showed a positive reaction only in the cytoplasmic protrusions released by the columnar cells, with the presence of a few positive granules (Fig. 4.7 C).

The other treatments (BRS 257, BR 16, Dowling) had no positive granules in their structures.

4 PHOTOGRAPHIC DOCUMENTATION

Presence of columnar cells (Co) with positive PAS granules (→); cytoplasmic protrusions (P) with positive PAS granules (→). Goblet cells (Ca) with positive PAS granules (→); regenerative cells (Cr) with positive PAS granules (→). Bars =10μm.

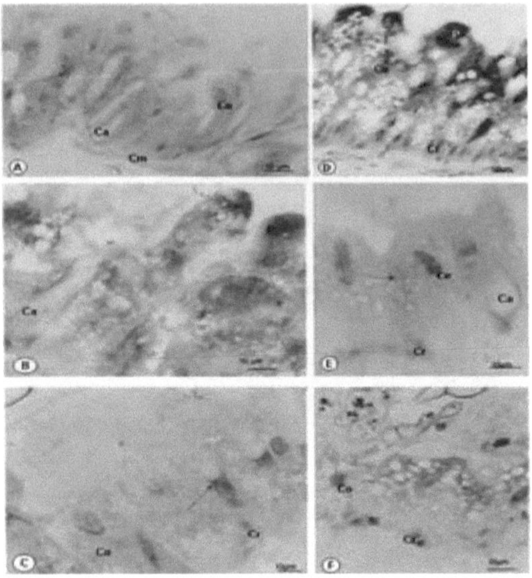

Figure 4.1 - Photomicrographs of the midgut *of A. gemmatalis* larvae susceptible to *Ag* MNPV treated with the following genotypes: BRS 257 (A); BR16 (B); Dowling (C), PI 229358 (D), IAC 100 (E) and PI 227687 (F). Histochemistry for neutral polysaccharides (PAS).

Presence of columnar cells (Co) with PAS-positive granules (→); goblet cells (Ca); regenerative cells (Cr). Bars = 10μm.

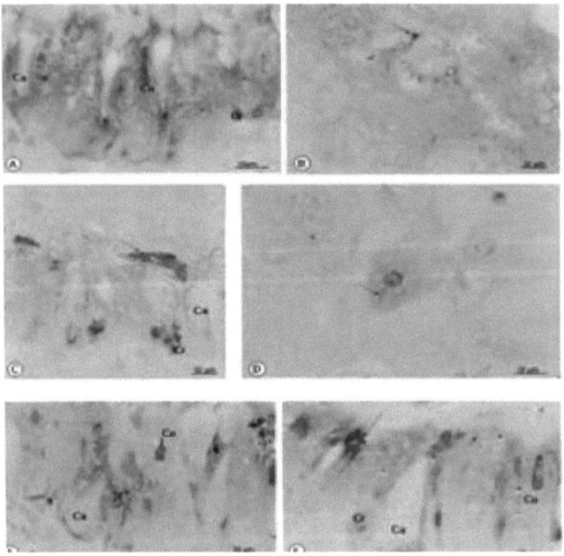

Figura 4.2 - Photomicrographs of the midgut of *AgMNPV-resistant A. gemmatalis* larvae treated

with the following genotypes: BRS 257 (A); BR16 (B); Dowling (C), PI 229358 (D), IAC 100 (E) and PI 227687 (F). Histochemistry for neutral polysaccharides (PAS).

Positive reaction to the Fast Green method in columnar cells (Co);

Goblet cells (Ca) and Muscle layer (Cm). Bars= 50µm.

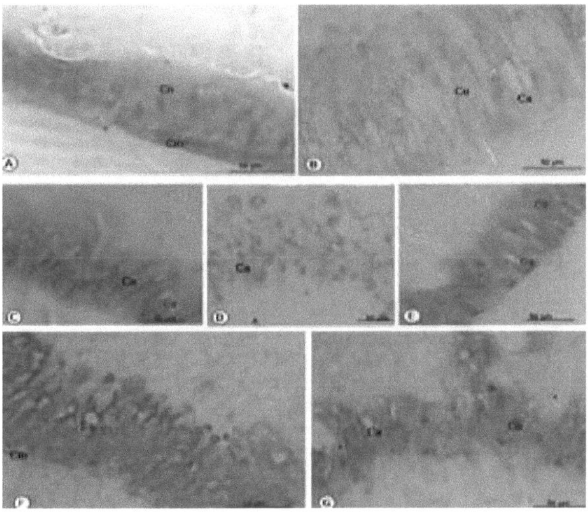

Figura 4.3 - Photomicrographs of the midgut *of A. gemmatalis* susceptible and resistant to *Ag* MNPV treated with the genotypes BRS 257 (A and G); BR16 (B); Dowling (C), PI 229358 (D), IAC 100 (E) and PI 227687 (F). Fast Green 0.1% total protein histochemistry.

Columnar cells (Co) with cytoplasmic protrusions (P) with nucleus (n); Goblet cells (Ca); Regenerative cells (Cr).Muscle layer (Cm) .Bars= 40µm

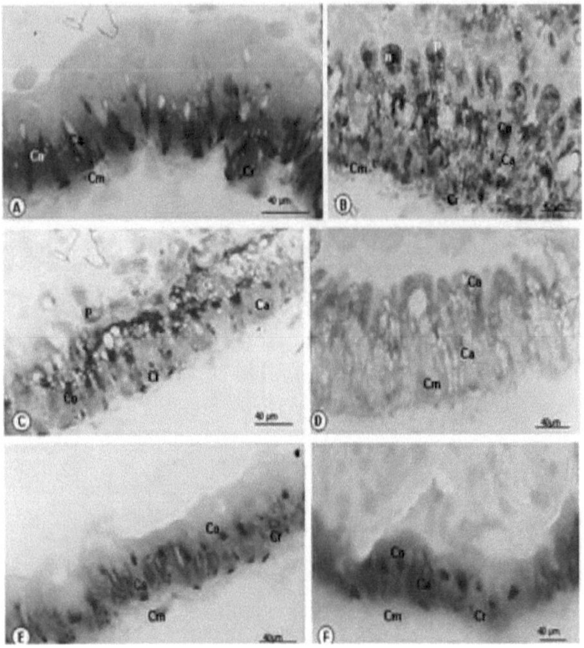

Figura 4.4 - Photomicrographs of the midgut *of A. gemmatalis* larvae susceptible to AgMNPV treated with the genotypes BRS 257 (A); BR16 (B); Dowling (C), PI 229358 (D), IAC 100 (E) and PI 227687 (F). Total protein histochemistry with bromophenol mercuric blue according to Pearse (1972)

Columnar cells (Co) Goblet cells (Ca); Regenerative cells (Cr). Bars= 40µm

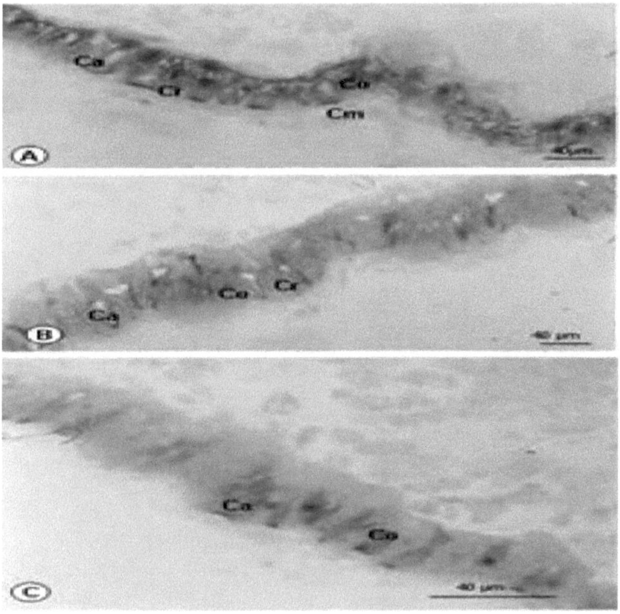

Figura 4.5 - Photomicrographs of the midgut of *AgMNPV-resistant A. gemmatalis* treated with the genotypes BRS 257 (A), IAC 100 (B) and PI 227687 (C). Histochemistry for the detection of total proteins with bromophenol mercuric blue according to Pearse (1972)

Columnar cells (Co); goblet cells (Ca); and muscle layer (Cm). Bars= 40µm.

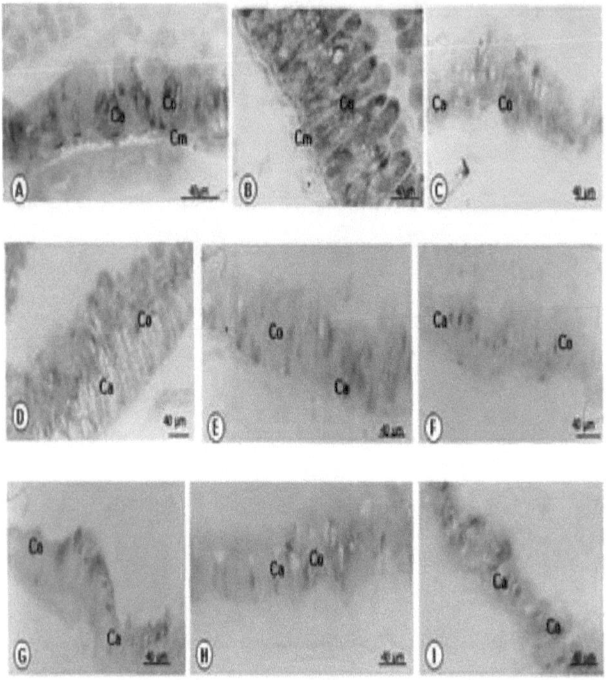

Figura 4.6 - Photomicrographs of the midgut *of A. gemmatalis* susceptible and resistant to *Ag* MNPV treated with the genotypes BRS 257 (A and G), BR 16 (B); Dowling (C); PI 229358 (D); IAC 100 (E and H) and PI 227687 (F and I). Histochemistry for total protein detection with bromophenol blue according to Coello (1989)

Columnar cells (Co) with cytoplasmic protrusions (P); goblet cells (Ca) and muscle layer (Cm). Bars= 10µm.

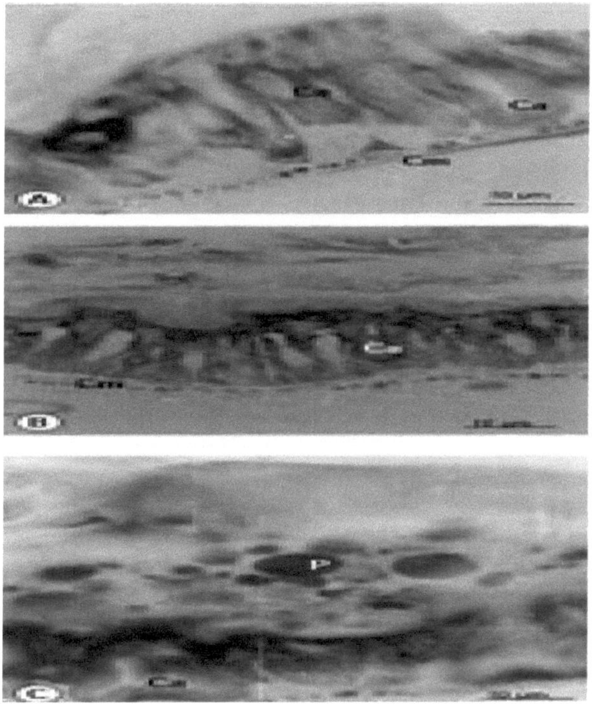

Figura 4.7 - Photomicrographs of the midgut of *AgMNPV-resistant A. gemmatalis* treated with the genotypes PI 229358 (A); IAC 100 (B) and PI 227687 (C). Histochemical detection of lipids according to Hernandez-Blazquez et al (1989).

5 DISCUSSION

The epithelial layer of the midgut *of Anticarsia gemmatalis* larvae susceptible and resistant to *AgMNPV* showed different levels of reaction intensity to the proposed histochemical methods in the three cell types, varying according to the treatment.

Columnar cells were the most positive to the Periodic Acid-Schiff (PAS) method, with the presence of PAS+ granules irregularly distributed throughout the cell cytoplasm. In both larval populations, the treatment with the BRS 257 genotype showed the highest intensity of PAS-positive granules. The IAC 100 and PI 227687 treatments showed strong positivity, but only in resistant larvae, while in LS the IMs showed moderate and weak positivity respectively. Another component of the epithelial layer that showed an intense concentration of PAS-positive granules were the cytoplasmic protrusions of susceptible larvae treated with the PI 229358 genotype and moderately positive for the BR16 genotype, and in resistant larvae after treatment with the PI 227687 genotype.

The presence of PAS-positive granules in the columnar cells and protrusions of the midgut of susceptible and resistant *A. gemmatalis* larvae is corroborated by the results found by some authors who identified these structures in the columnar cells and cytoplasmic protrusions of the midgut in larvae of other lepidopterans such as *Diatrea saccharalis* (PINHEIRO et al., 2008); *Spodoptera frugiperda* (CORREIA et al., 2009); Alabama argilacea (SOUZA et al., 2009).

The identification of this material is not restricted to the Lepidoptera order, and it has been identified in other insect orders and species such as *Triatoma infestans* (Hemiptera-Heteroptera) reported by Burgos; Gutierrez (1976); *Dermatobia hominis* and *Aedes aegypti*, (Diptera) by Lello; Vieira, 2001; Arruda; Oliveira, Silva (2003) and *Tropidacris collaris* Orthoptera Wanderley-Teixeira (2006).

According to Terra and Ferreira (1991), the functions of these polysaccharides found in columnar cells are to absorb digestive products and produce enzymes. They can vary in concentration according to the region of the midgut, with the highest concentration being found in the distal region of the midgut (RIBEIRO et al., 1997; PINHEIRO et al., 2008).

In goblet cells, this reaction is rarely found, and when it is localized, it occurs mainly in the cytoplasm of these cells. In our work, the reaction occurred only in the goblet cells of the midgut of resistant larvae treated with the BRS 257 genotype and susceptible larvae treated with the PI 229358 genotype. Pinheiro et al. (2008) identified PAS+ granules in this same cell type in *Diatrea saccharalis* larvae and suggested, like other authors, that the presence of polysaccharides may be related to the activity of these cells in assisting columnar cells such as absorption of metabolites and ion homeostasis (MOFFET et al., 1995; LEHANE; BILLINGSLEY, 1996; CHAPMAN, 1998).

The regenerative cells showed PAS+ material only in the PI 229358 treatment for susceptible larvae. Lehane and Billingsley (1996) reported that because these cells are involved in the epithelial renewal process, they would present PAS-positive material. This fact was also reported in *Aedes aegypti* larvae, in which they showed intense activity in epithelial renewal (ARRUDA; OLIVEIRA; SILVA, 2003).

A positive PAS reaction indicates the presence of neutral polysaccharides, which may represent deposits of glycogen and other neutral sugars. Glycogen deposits were identified after diastase, in which the slides containing the MI sections and submitted to PAS showed a reduction in some treatments in relation to the intensity of the PAS-positive granules. The BRS 257 (control), BR16 and Dowling treatments only showed a reduction in LR, while the intensity was the same for LS. We can therefore infer that the LS treated with these genotypes had neutral polysaccharides in their cells, but not glycogen.

The IAC 100 treatment was the only one to show a decrease in positivity in both larval types, indicating that these larvae had a large amount of glycogen in their cells. The other treatments showed specific changes in the midgut, varying according to the treatment and the reaction structure.

According to Kilby (1965) glycogen and other carbohydrates are important in metabolism, as they are considered an important reserve that can be mobilized and used in situations that require energy expenditure by the insect, such as moulting (metamorphosis), reproduction, defence, among other factors.

Hock, Pilarska and Dobart studied larvae *of Lymantria dispar* (Lepidoptera: Lymantriidae) found that the concentration of glycogen increased with the level of larval development in both healthy larvae and those infected with the entomopathogen *Endoreticulatus schubergi*. However, after a few days of infection, this concentration of glycogen decreased drastically in the infected larvae and the healthy ones showed higher levels of glycogen concentration. Our results indicate that the polysaccharide present in the cytoplasm of the midgut cells of susceptible larvae is not an energy polysaccharide, such as glycogen, but a structural polysaccharide, which could even be chitin, a structural polysaccharide that is very common in arthropods.

Among the polysaccharides there are other groups which are called acidic and which can be classified as highly sulphated acidic polysaccharides, sulphated acidic polysaccharides and acidic polysaccharides. The midgut of the susceptible and resistant larvae, after being subjected to the histochemical process of toluidine blue at different pHs, showed no reaction to the method, because according to Pearse (1972) the presence of these acids would be evidenced by blue or pink coloration, and the MI of these larvae showed no reaction, indicating the absence of acidic carbohydrates.

In addition to polysaccharides, other substances are present in insects, such as proteins which have various functions in insects, such as the production of silk threads or venom glands (CHAPMAN, 1998).

Dean et al. (1985) reported that proteins are present in the vast majority of insect orders, except in Hymenoptera, which has no reports of the presence of protein granules.

In our studies, the Fast Green histochemical reaction for proteins in the midgut of susceptible *A. gemmatalis* larvae was positive, with the presence of protein granules in the goblet cells of all the treatments, in the muscle layer (BRS257, Dowling and PI 227687) and in the columnar cells (Dowling and PI 227687). In resistant larvae, only the control (BRS 257) showed granules in most structures, while the PI 227687 genotype showed a reaction in goblet cells. Our results are corroborated by data from Lehninger (2006), who divides proteins into two classes: fibrous (structural functions in cells and tissues), as described in our results, and globular (physiological function).

With Pearse's (1972) method for detecting total proteins using bromophenol mercuric blue, we observed the presence of proteins in the midgut epithelium of susceptible larvae in all treatments, and in resistant larvae, some treatments were positive to the method applied, such as BRS 257, Dowling, IAC 100 and PI 227687. According to Pearse (1972) the reaction shows most of the proteins colored light blue with a slight tendency towards green, which was also observed by us in the cell nuclei, muscle layer and cytoplasmic protrusions in both susceptible and resistant larvae, varying in the intensity of the coloration.

The cytoplasmic protrusions generally showed columnar cell nuclei, hence the strong reaction.

The presence of proteins may be related to various factors such as insect ecdysis, a source of energy reserves, especially during periods of food shortage, tissue formation and injuries (DEAN et al., 1985).

Rocha (2009) in a study with *Triatoma vitticeps* observed an accumulation of proteins in the digestive cells of the midgut of these insects when they were "food fasting", and the more days they were fasting, the higher the concentration of these proteins. This may be related to the injuries suffered and the accumulation of proteins in the body that will be converted to maintain the insect's organs.

Considering that insects use various molecules accumulated in their organism in evolutionary processes (metamorphosis) as well as in the protection and maintenance of their organs, lipids are substances that help in these processes.

Lipids have the characteristic of having a higher energy value than carbohydrates, thus becoming energy sources that can be accumulated in the insects' bodies to be used during their development.

The histochemical reaction for the determination of lipids in the midgut epithelium of susceptible and resistant *A. gemmatalis* larvae showed the presence of *a* few drops of lipids only in resistant larvae from the IAC 100, PI 229358 and PI 227687 treatments; the other treatments did not show a positive reaction to the method applied.

Pinheiro (2001), analyzing the midgut regions of *Diatrea saccharalis*, observed the presence of lipid droplets in these three regions and in all epithelial cells. In *A. gemmatalis*, higher concentrations of lipids are found in the fat body, as reported by Carvalho, (2008), and the availability of this reserve occurs in the pupal stage. This fact was also observed in *Lymantria dispar* larvae, which showed higher lipid levels in fifth and sixth instar larvae in both healthy and virus-infected larvae.

6 CONCLUSIONS

The histochemical techniques applied showed that the midgut of *Anticarsia gemmatalis* larvae susceptible and resistant to *AgMNPV* had compounds such as neutral polysaccharides, glycogen, proteins and lipids in their structures. These compounds play different roles, both in tissue maintenance and in cell physiology, especially in epithelial cells:

With regard to neutral polysaccharides, the midgut of susceptible and resistant larvae, when compared to the control, showed greater positivity to PAS in the IAC 100 treatments, and in PI 227687 in resistant larvae, mainly in the columnar cells, and in these cells, the greatest participation, for the IAC 100 genotype, was of reserve polysaccharides, in this case glycogen, while in the other genotypes the presence was of structural polysaccharides.

Total proteins were intensely marked in columnar cells by fast green 0.1% in all treatments in susceptible larvae, while resistant larvae showed no reaction in all treatments. Using the mercuric bromophenol blue method, the treatments with the highest concentrations of proteins were BR 16, Dowling and IAC 100, with reactions varying between all the structures of the intestinal epithelium, while in resistant larvae the best results were seen in IAC 100 and PI 227687. By the bromophenol blue histochemical reaction, the PI 229358 genotype showed a higher concentration of proteins compared to the other treatments, and in the resistant larvae the IAC 100 treatment showed the highest positive reaction. Thus, according to the techniques used, the treatment with the IAC 100 genotype, both in larvae susceptible and resistant to *AgMNPV*, was the one that showed the highest concentration of proteins in the epithelial layer of the midgut, especially in the goblet cells.

With regard to lipids, the resistant larvae showed lipid droplets when treated with the PI 229358, IAC 100 and PI 227687 genotypes.

Thus, the presence of these chemicals indicates a response by the insect's physiological system to the action of the toxic substances in the midgut.

REFERENCES

ABOT, A.R.; et al. Development of resistance by *Anticarsia gemmatalis* from Brazil and the United States to a nuclear polyhedrosis virus under laboratory selection pressure. **Biological Control**, v.7, n. 1, p.126-130, aug. 1996.

APPEL, H.M. Phenolics in ecological interactions: the importance of oxidation.

Journal of Chemical Ecology, v.19, n.7, p.1521-1551, jul.1993.

ARRUDA, W.; OLIVEIRA, G.M.; SILVA, I.G. Toxicity of the ethanolic extract of *Magonia pubescens* on *Aedes aegypti* larvae.**Revista da Sociedade Brasileira de Medicina Tropical**, v.36, n.1, p.17-25, jan-feb. 2003.

BILLINGSLEY, P.F.; LEHANE, M.J. Structure and ultrastructure of the insect midgut.

In: LEHANE, M.J., BILLINGSLEY, P.F. (Eds). **Biology of the Insect Midgut**.

London: Chapman; Hall, 1996, p.3-30.

BINDER, B.F.; BOWERS, W.S. Precocene II-induced changes in the anatomy of midgut goblet cells of last instar larvae *of Helicoverpa zea* (Boddie) (Lepidoptera: Noctuidae). **International Journal of Insect Morphology and Embryology**. Oxford v. 23, n.2, p.127-134, Apr. 1994.

BROWN, M.R., RAIKEL, A.S., LEA, A.D. Ultrastructure of midgut endocrine cells in the adult mosquito, *Aedes aegypti*. **Tissue and Cell**, Ouxford v.17, n.5, p. 709-722, aug. 1985.

BURGOS, M.H.; GUTIÉRREZ, L.S. The intestine of *Triatoma infestans*. **Journal of Ultrastructure Research**, v.57, p.1-9, 1976.

CARVALHO, R.B.R. **Histological, histochemical and ultrastructural study of the fat body of *Anticarsia gemmatalis* (Hübner, 1818) (Lepidoptera: Noctuidae) larvae**. 2008. Dissertation (Master's Degree in Agronomy).Universidade Estadual de Londrina - UEL, Londrina

CAVALCANTE, V.M.; CRUZ-LANDIM, C. Types of cells present in the midgut of the insects: a review. **Naturalia**, Rio Claro, v.24, p. 19-40, 1999.

CERRI, P., SASSO-CERRI, E. Staining methods applied to glycol methacrylate embedded tissue sections. **Micron**, v.34, n.8, p.365, 2003.

CHAPMAN, R.F. **The Insects: structure and function**. 4.ed. Cambridge: Harvard University Press, 1998.

CHIANG, A.S.; YEN, D.F.; PENG, W.K. Defense reaction of midgut epithelial cells in the rice moth larva (*Corcyra cephalonica*) infected with *Bacillus thuringiensis*.

Journal Invertebrate Pathology, v.47,n.3, p.333-339, May 1986.

COAST, G.M., KAY, I. The effects of *Acheta* diuretic peptide on isolated Malpighian tubules from the house cricket, *Acheta domesticus*. **The Journal of Experimental Biology**, London, v.187, n.1, p.225-243, feb. 1994.

COELLO, S.A. A new staining schedule for formalin-fixed, glycol methacrylate- embedded fish ovaries. **Journal Fish Biology**, v.34, p.195-213, 1989.

CORREIA, A.A. Morphology of the alimentary canal of caterpillars *of Spodoptera frugiperda* (J.E. Smith) (Lepidoptera: Noctuidae) fed with leaves treated with Nim. **Neotropical Entomology**, Vacaria, v.38, n.1, p.83-91, jan-feb. 2009.

DEAN ,R.L. et al., Structure of fat body. In: KERBUT, G.A.; GILBERT, L.I. (Ed.). **Comprehensive insect physiology, biochemistry and pharmacology**. Oxford: Pergamon Press, 1985, p.155-210, v.3

DE PRIESTER, W. Ultrastructure of the midgut epithelial cells in the fly *Callifora erythrocephala*. **Journal of Ultrastructure Research.** New York v.36, n.5, p. 783805, set. 1971.

EISEMANN, C.H.; BINNINGTON, K.C. The peritrophic membrane : its formation, structure, chemical composition and permeability in relation to vaccination against ectoparasitic arthropods. **International Journal for Parasitology**, London, v.24, n.1,p.15-26, feb. 1994.

FELTON, G.W. et al. Activation of plant foliar oxidases by insect feeding reduces nutritive quality of foliage for noctuid herbivores. **Journal of Chemical Ecology**, California, v.15, n.12, p.2667-2694, 1989.

FELTON, G.W.; DUFFEY, S.S. Enzimatic antinutritive defenses of the tomato plant against insects.In: HEDIN, P. (Ed.) **Naturally occurring pest bioregulators.**

Washington: American Chemical Society, 1991. p.166-197.

FIORE-DONNO, G.; BAUME, L.J. Étude histochimique de la dentinogenese humaine. Helvetica ondontologica acta, v. 10,Supplementum IV, 1996.

FLIPSEN, J.T.M.; et al. Passage of *Autographa californica* nuclear polyhedrosis virus through the midgut epithelium of *Spodoptera exigua* larvae. **Virology**, New York v.208, p.328-335, 1995.

GAZZONI, D. L.; et al. **Soybean pest management**. Londrina: Embrapa-CNPSo, 1988.

GAZZONI, D.L. HÜLSMEYER,A.; HOFFMANN-CAMPO, C.B. Effect of different doses of rutin and quercetin on the biology of *Anticarsia gemmatalis* Hübner, 1818 (Lep., Noctuidae). **Revista Agropecuaria Brasileira**, Brasilia, v.32, n.7, p. 673-681, jun./jul. 1997.

GRANADOS, R.R.; LAWLER, K.A. In vivo pathway of *Autographa californica* baculovirus invasion and infection. **Virology**, New York, v.108, n.30, p.297-308, jan. 1981.

GRANADOS, R.R.; WILLIAMS, K.A. In vivo infection and replication of baculoviruses. In: GRANADOS, R.R.; FEDERICI, B.A. (Eds.). **The biology of baculoviruses**. Boca Raton: CRC, 1986, p.89-108 v.1.

HARBONE, J.B. Flavonoid pigments. In: ROSENTHAL, G.A.; BEREMBAUM, M.R. (Eds.). **Herbivores and their interaction with secondary metabolites**. London: Academic Press, 1991, p.389-429.

HERNANDEZ-BLAZQUEZ, F.C.; et al., Fat intestinal absorption in the catfish A histochemical study in glycol methacrylate embedded tissue. **Gegenbaur morphology Jahrb. Leipzig,** v.135 , p.941-946, 1989.

HOCH,G.; PILARSKA, D.K.; DOBART, N. Effect of midgut infection with the microsporidium *Endoreciculatus schubergi* on carbohydrate and lipid levels in *Lymantria dispar* larvae. **Journal Pest Sicence**, v.82, p. 351-356, jun. 2009.

HOFFMANN-CAMPO,C.B.; et al. Mechanisms of resistance of soybean genotypes: non-preference test for *Anticarsia gemmatalis* Hübner, 1818 (LepzNoctuidae). **Pesquisa Agropecuâria Brasileira**, Brasilia, v.29, n.4, p. 513-519, abr. 1994.

HOFFMANN-CAMPO, C.B.; **Role of the flavonoids in the natural resistance of soyabean to *Heliothis virescens* (F.) and *Trichoplusia ni* (Hübner)**. 1995.

Dissertation (PhD in Botany) - 165 p. The University of Reading, Reading.

HOFFMANN-CAMPO, C.B.; et al. **Soybean pests in Brazil and their integrated management.** Londrina: EMBRAPA-CNPSo, 2000.

HOFFMANN-CAMPO, C.B.; et al. Pre-ingestive and and post-ingestive effects of soybean extracts and rutin on *Trichoplusia ni* growth. **Entomologia Experimentalis et Applicata**, Amsterdam, v.98, n.2, p.181-194, feb. 2001.

HOFFMANN-CAMPO, C.B.; et al. Detrimental effect of rutin on *Anticarsia gemmatalis*. **Pesquisa Agropecuaria Brasileira**, Brasilia, v.41, n.10, p. 1453-1459, Oct. 2006.

KAWANISHI, C.Y.; et al. Entry of insect virus *in vivo* by fusion of viral envelope and microvilli membrane. **Journal of Invertebrate Pathology,** v.20, p.104-108, jul. 1972.

KILBY, B.A. Intermediary metabolism and the insect fat body. In: GOODWIN, T.W. **Aspects of insect biochemistry**. New York: Academic Press, 1965. p.39-48.

KOGAN, M. Natural chemicals in plant resistance to insects. **Journal of Research**, Iowa State v.60, p.501-527, 1986.

KUBO, I.; HANKE, F.G. Chemical methods for isolating and identifying phytochemicals biologically active in insects. In: MILLER, J.R.; MILLER, T.A. (Eds) **Insects plants interactions**. New York: Spring Verlog. 1986 p.225-249.

LARA, F.M. **Principios de resistência de plantas a insetos**. Sao Paulo: icone, 1991.

LEHANE, M. J.; BILLINGSLEY, P. F. (Eds.). **Biology of the insect midgut.** London: Chapman; Hall, 1996.

LEHNINGER, A. L.; NELSON, D.L. **Lehninger Principles of Biochemistry**. Sao Paulo: Sarvier, 2006

LELLO, E ; VIEIRA, A.M. Post-embryonic development of the midgut *of Dermatobia hominis* (Linnaeus Jr.) (Dipatera:Cuterebridae). **Revista Brasileira de Zoologia**, v.18, p.91-98, 2001.

LEVY, S.M. *et al.* The larval midgut of *Anticarsia gemmatalis* (Hübner) (Lepidoptera: Noctuidae): light and electron microscopy studies of the epithelial cells. **Brazilian Journal of Biolpogy**, Sao Paulo, v.64, n. 3b, p.1-6, aug. 2004.

LEVY, S.M.;et al. A morphometric study of the midgut in resistant and non-resistant Anticarsia gemmatalis (Hübner) (Lepidoptera: Noctuidae) larvae to its nucleopolyhedrovirus (AgMNPV). **Journal of Invertebrate Pathology**, v.101, n.1, p.17-22, Apr. 2009.

LIMA, L.R.P. et al. Determination of lipase activity in the presence of morin, naringenin, naringin and rutin. **Revista de Ciência e Agrotecnologia**, Lavras, v. 23, n. 3, p. 626-631, jul./set. 1999.

LISON, L. **Histochemie et Cytochemie Animales, Principes et Méthods**. Paris: Gauthier-Villars, 1960, v.2.

LOPES, R.M. et al. Effect of different doses of rutin on the concentrations of albumin, creatine, urea and uric acid in the serum of rabbits. **Revista Brasileira de Farmacia**, Rio de Janeiro, v. 86, n.1, p. 35-38, jan. 2005.

MCMANUS, J.F.A.; CASON, J.F. Carbohydrates Histochemistry Studied by Acetylations Techniques. I Periodic Acid Methods. **The Journal of Experimental . Medicine**. Rockefeller, v.91, p. 651-654, 1950.

MELLO, M.L.S., VIDAL, B.C. **Praticas em biologia celular**. Sao Paulo: Edgar Blücher, 1980.

MOFETT, F.; KOCH, A. Electrophysiology of K^+ transport by midgut epithelium of lepidopteran insect larvae. **The Journal of Experimental Biology**, London, v.198, n.10, p.2115-2125,Oct. 1995.

MONNERAT, R.G.; BRAVO, A. Bioinsecticidal proteins produced by the bacterium *Bacillus thuringiensis* mode of action and resistance. In: MELO, I.S.; AZEVEDO, J.L. (Ed). **Biological control**. Jaguariùna. Embrapa-CNPMA, 2000, v.3, p.163-200.

MOSCARDI, F.; CARVALHO, R.C.Z. Consumption and utilization of soybeans by *Anticarsia gemmatalis* Hüb. (Lepidoptera: Noctuidae) infected at different larval stages by its nuclear polyhedrosis virus. **Anais Sociedade Entomològica do Brasil**, Londrina, v.22, n.2, p.267-280, jun.1993.

OMYILAGHA, J.C.; GROTEWOLD, E. The biology and structural distribution of surface flavonoids. **Recent Research developments in plant science**, Columbus Ohio, v.2, p. 53-71, 2004.

PANIZZI, A.R. Integrated management of soybean pests in Brazil. In: CROCOMO, W.B. (Ed.). **Integrated pest management.** Sao Paulo: UNESP-FUNEP, 1990. p.293-321.

PATEL, N.; HAYES, T.K., COAST, G.M. Evidence for the hormonal function of a CRF-related diuretic peptide (*Locusta-DP*) in *Locusta migratoria*. **The Journal of Experimental Biology,** London, v.198, n. 3, p.793-804, mar. 1995.

PEARSE, A.G.E. **Histochemisrtry:** theoretical and applied. Edinburgh: Churcill Livinsgtone, 1972.

PINHEIRO, D.O. **Morphology of epithelial cells in different regions of the mesenterium of larvae *of Diatrea saccharalis*, Fabricius 1794 (Lepidoptera: Pyralidae)**. 2001 Monograph (Bachelor's Degree in Biological Sciences) Instituto de Biociências da Universidade Estadual Paulista - UNESP Campus de Botucatu, Botucatu.

PINHEIRO, D.O. et al., Morphological Regional Differences of Epithelial Cells along the Midgut in Diatraea saccharalis Fabricius (Lepidoptera: Crambidae) Larvae.

Neotropical Entomology, Vacaria, v.37, n.4, p.413-419, jun./ago. 2008.

PIUBELLI, G.C. **Bioactivity of soybean genotypes resistant to *A. gemmatalis* Hübner (Lepidoptera: Noctuidae) and interactions of their chemical substances with natural enemies**. 2004.Thesis (Doctorate in Biological Sciences - Entomology), Federal University of Paranà, Curitiba, Paranà. 152p.

PIUBELLI, G.C et al. Are chemical compounds important for soybean resistance to *Anticarsia gemmatalis* ? **Journal of Chemical Ecology**, California, v.31, n.7,p.1509- 1525, jul. 2005.

PIUBELLI, G.C. et al. Baculovirus-resistant *Anticarsia gemmatalis* response differently to dietary rutin. **Entomologia Experimentalis et Applicata**, v.119, p.5360, 2006.

PRATT, G.E. et al. Identification of allatostatin from adult *Diploptera punctata*. Biochemical and biophysical research communications, United States, v.163, n.3, p. 1243-1247, sep. 1989.

RIBEIRO A F, FERREIRA C, TERRA, W R Morphological basis of digestion in insects. **Acta Microscopica** v.6: p.71-73. 1997.

RICHARDS, O.W., DAVIES, R.G. **Imms' general textbook of entomology**. 10. ed. London: Chapman; Hall, 1994. v.1.

SALVADOR, M.C. **Effect of soybean genotypes and flavonoids on the biology and midgut of *Anticarsia gemmatalis*.** *2008.* Dissertation (Master's Degree in Agricultural Entomology). Faculty of Agricultural and Veterinary Sciences - UNESP Jaboticabal Campus, Jaboticabal.

SCOOFS, L., VAN DEN BROECK, J., DE LOOF, A. The myotropic peptides of *Locusta migratoria*: structures, distribution, fuctions and receptors. **Insect biochemistry and molecular biology**. England, v.23, n.8, p.859-81, dec. 1993.

SILVA, R.R. et al. Hypolipidemic effect of the flavonoids naringin and rutin. **Archivos Latinoamericanos de Nutrición**, Caracas, v.51, n.3, p. 258-264, Apr. 2001.

SILVA, R.R. et al. Effect of flavonoids on arachidonic acid metabolism.

Medicina, Ribeirao preto, v.35, p. 127-133, abr./jun. 2002.

SOUSA, M.E. et al. Ultrasctructure of the *Alabama argilacea* (Hübner) (Lepidoptera:Noctuidae) midgut. **Micron**, v.40, p.743-749, 2009.

STAMP, N.E.; SKROBOLA, K.L. Failure to avoid rutin diets results in altered food utilizatin and reduced growth rate of *Manduca sexta* larvae. **Entomologia Experimentalis et Applicata**, Netherlands, v. 68, n.2, p.127-142, aug. 1993.

TERRA, W.R.; FERREIRA, C. The digestion of insects. **Ciência Hoje,** v.12, p.28-38, 1991.

TERRA, W. R. The origin and functions of the insect peritrophic membrane and peritrophic gel. **Archives of Insect Biochemistry and Physiology**, New York, v.47, n. 2, p.47-61, jun. 2001.

TOLEDO, T., et al., Effects of differents rutin dosages on chloride, calcium and phosphorum levels serum of rabbits. **Revista Chilena de Nutrición**, v.30,n.3, p.263270, dec. 2003.

VOLKMAN, L.E. Nucleopolyhedrovirus interactions with their insect hosts. **Adv. Virus Res.**, v.48, p.313-348, 1997.

WANDERLEY-TEIXEIRA, V. et al., Histological description of the midgut and the pyloric valve of *Tropidacris collaris* (Stoll, 1813) (Orthoptera: Romaleidae). **Brazilian Journal of Biology**, v.66, n.4, p.1045-1049, nov. 2006.

WANG, P.; GRANADOS, R.R. Molecular structure of the peritrophic membrane (PM): identification of potential PM target sites for insect control. **Archives of Insect Biochemistry and Physiology**, New York, v.47, n.2, p.110-118, jun. 2001.

WOODHEAD, A.T.; et al. Primary structure of four allatostatins: neuropeptide inhibitors of juvenile hormone synthesis. **Proceedings of the National Academy of Sciences of the United States of America**, Iowa City, v.86, n. 15, p.5997-6001, Aug. 1989.

8 GENERAL CONCLUSIONS

The midgut of *Anticarsia gemmatalis* larvae susceptible and resistant to *Ag* MNPV showed changes in the cells of the epithelial layer after treatment with soybean genotypes composed of flavonoids (rutin and genistein) at different concentrations.

These morphological changes were specific and occurred mainly in the columnar cells, which showed intense vacuolization and the release of cytoplasmic protrusions that sometimes contained cell nuclei into the lumen. Other structures such as the peritrophic membrane and the muscular layer also showed some alterations in their structures depending on the genotype.

The damage caused to susceptible larvae was promoted mainly by the PI 227687 genotype, while in resistant larvae these alterations were caused mainly by the IAC 100 and PI 227687 genotypes, which caused total destruction of the epithelial layer of the insects' intestines.

The epithelium showed superficial morphological changes, in which three distinct regions could be seen: the anterior region with columnar cells with varied surfaces and rare cytoplasmic protrusions; the middle region with some cytoplasmic projections and varied columnar cell surfaces; and the posterior region with a large number of cytoplasmic projections that varied in shape.

These projections were found in greater quantity in the midgut of both larval populations treated with Dowling and PI 227687 genotypes, thus showing that these structures are related to the elimination of toxic substances or injured cells, since these genotypes have rutin and genistein in high concentrations.

Histochemical analysis showed that LS and LR are composed of polysaccharides, glycogen, proteins and lipids, and that they perform various functions in the insect digestive system.

These compounds are considered to be sources of energy reserves. The IAC 100 and PI 227687 genotypes showed higher concentrations of chemical compounds in the midgut cells of susceptible and resistant larvae.

In general, the IAC 100 and PI 227687 genotypes are recommended for planting because they have pest-resistant characteristics that promote intense changes in the midgut of *A. gemmatalis* larvae in both populations,

MIX
Papier aus verantwortungsvollen Quellen
Paper from responsible sources
FSC® C105338

Printed by Books on Demand GmbH, Norderstedt / Germany